HOW TO PRONOUNCE DRUG NAMES

A Visual Approach to Preventing Medication Errors

Tony Guerra, Pharm.D.

2017

How to Pronounce Drug Names: A Visual Approach to Preventing Medication Errors

Tony Guerra, Pharm.D.

First Edition

ISBN: 978-1-365-56630-1

To Mindy,

Brielle, Rianne, and Teagan

Author's note

The knowledgeable patient

Can you remember, pronounce, and spell the active ingredient in the last medicine you took or gave your child? If you're an administrator, can everyone in your facility who provides patients with medications spell and pronounce those medications correctly?

If you have any doubt—whether you're reading this book in a doctor's waiting room, a pharmacy, or a hospital—I wrote this book for you.

Call me Tony. I'm a parent, and I've been a pharmacist for 19 years. A revolution is happening in health care. In the past, nurses, pharmacists, and doctors spoke together at length with families about their concerns and medicines.

But now, patients receive only distracted attention as their health professionals tangle with billing, electronic records, and insurance obstacles. This shortened communication frustrates doctors and patients. Patients find themselves isolated because they don't speak the language of medicine.

Patients now rely heavily on self-care through the Internet, friends, and relatives. Those who decipher the medical language and become knowledgeable patients find satisfaction, and are less likely to complain or sue a doctor or hospital for malpractice. When patients complain, half the time it's because of communication. Many institutions, I believe, focus narrowly on improving health care workers as teachers without helping patients become more effective students. But, where should patients start their medical education? They can start by learning to turn medication names' text into speech, and speech into text.

Traditionally, the language of medicines — pharmacology — requires the study of anatomy, physiology, inorganic and organic chemistry, and biochemistry. This is the health professional's road. But, just as we build chemical names from chemical parts to form generic drug names, we can form generic names from common English words. With a little audio cut and paste magic, it's possible to hear the drug name in your own voice without all that extra education I mentioned. I'll teach you how to do that later in this book.

WHAT ENGLISH DIALECT DO *YOU* SPEAK?

Whether you speak an English you learned in the Americas, Europe, Australia, Africa, or Asia, it doesn't matter. By using plain English words, the generic names' sounds mirror those of your particular English. For example, pull up the Oxford English Dictionary at the oed.com website. Look up the phonetic pronunciation of the analgesic, "ibuprofen, i-b-u-p-r-o-f-e-n." Now, in the OED, look up the phonetic pronunciations for: the letter "i;" the "b-u, bu" from "bureau;" the "p-r-o, pro," from "proton;" and the "f-e-n, fen," from "fender." They match.

Do the same for ibuprofen from Merriam-Webster.com, an American dictionary. You'll see that the sounds match again. Sure, the sounds are different, but so is the way Americans pronounce ibuprofen; but the individual sounds often mirror the English words of the country of origin. Cool, right? Generic medication names are a welcome universal language in a health care system with many parts that don't often connect.

WHY ARE GENERIC DRUG NAMES BETTER?

Generic names provide three important advantages.

1. Generic names foster interoperability, the ability to communicate in different health care settings. The generic name ibuprofen is the same at the corner drugstore, retail store, grocery store, or hospital,

and in the United States, United Kingdom, or Australia and beyond. The generic ibuprofen name unifies. Brand names Advil and Motrin in the U.S. or Nurofen in other parts of the world don't.

2. Generic names are precise. Drugs have a single generic name, but can have multiple brand names. Watch a pharmacist talk to a patient about an over-the-counter medicine. Good pharmacists always flip the box around to make sure they are looking at the generic names, the active ingredients.

3. Generic names are systematic. Brand names lack a taxonomy, a classification system. In contrast, generic names have meaningful prefixes, infixes, and suffixes. This helps you translate the drug in your hand to its class and related drugs. Alternative medications might cost much less.

Failing to absorb medication names results in frustration and danger. Learning them eases medical isolation, and puts you on the road to becoming a knowledgeable patient. You *can* engage in better self-care. First, let's clarify one problem that often threatens children, but which using generic names solves, and I'll do this with a short story.

WHAT ARE BRAND NAME EXTENSIONS?

I found an Internet request where a mom asked if it was okay that she gave Pepto Bismol liquid to her child. The answer is no. Pepto Bismol liquid is bismuth subsalicylate, a relative of aspirin. Parents should not give this to children because it can cause Reye's Syndrome, a serious neurologic disorder. She meant to give Pepto Children's, which contains calcium carbonate tablets, an antacid, and the active ingredient in Tums. But we shouldn't vilify this parent because she expected that a pink liquid should be correct for a child. How could she know childrens' tablets exist?

Brand names, like Pepto Bismol and Pepto Children's, are names drug manufacturers make to describe their version of a product. The technical phrase for the manufacturer's well-known and popular original name with different active ingredients is a "brand name extension." While brand name extensions historically cause confusion, it is easier as a patient to speak by using generic names than to hope this brand extension practice will change. When you send someone to the store for a medicine, ask for the active ingredient, the generic name, like bismuth subsalicylate or calcium carbonate. But how do we pronounce, spell, and remember words that bring up no images?

BRAND VERSUS GENERIC NAMES. IT'S ABOUT USABILITY.

We don't use generic names except with the most common drugs, like ibuprofen and acetaminophen. Why? Brand names have better usability.

Usability describes how easy something is to use. Pepto, P-e-p-t-o, is a staple in medicine cabinets. It's likely familiar to both the patient and the clerk at the grocery store who takes him to aisle eleven, second shelf from the bottom. Usability is good for "Pepto," as it's easy to say *and* recognizable. I have three kids, and I use Pepto all the time for indigestion. I'm not condemning the product here — I'm clarifying the active ingredient.

The two generic names, bismuth subsalicylate and calcium carbonate, combine for thirteen syllables. This provides poor usability and comprehensibility. This is disappointing because generic names are far better with interoperability.

GENERIC NAME INTEROPERABILITY

In the book, *Memorizing Pharmacology: A Relaxed Approach,* I wrote about drug stems — that is, parts of most newer *generic* medications which have a meaningful stem in the beginning (prefix), middle (infix), or end (suffix) position. In this case, the –tidine, or the six letters t-i-d-i-n-e in the suffix position, means that these four medications block histamine-2, resulting in less gastric acid. You can see below how the four generic names — cimetidine, famotidine, nizatidine, and ranitidine — are similar.

Generic name	Brand name	Generic name	Brand name
Cimetidine	Tagamet	Ci me ti dine	Ta ga met
Famotidine	Pepcid	Fa mo ti dine	Pep cid
Nizatidine	Axid	Ni za ti dine	Ax id
Ranitidine	Zantac	Ra ni ti dine	Zan tac

There is no such pattern in the brand names; rather, brand names may allude to the more general medication role or take parts of the brand name. For example

Tagamet. The "t-a-g" in "an<u>tag</u>onist" and "m-e-t" in "ci<u>met</u>idine"

Pepcid. The "p-e-p" in "<u>pep</u>tic" and "c-i-d" from "a<u>cid</u>"

Axid. Add "k" to acid's phonetic pronunciation, "a-s-i-d, a-sid," as "a-k-s-i-d, ak-sid," to make "a-x-i-d, Axid"

Zantac. Includes the "a-c" from "a<u>c</u>id" and "a-n-t" from "<u>ant</u>agonist"

To summarize, brand names currently have better usability than generic names, but are individualistic, imprecise, and unsystematic — which can lead to errors.

A PUZZLE WORTH SOLVING

Our puzzle, one I believe is worth solving, is to create a system that makes generic medication names more usable. This way, patients and parents can use the three advantages of generic names to protect themselves and those they love.

Generic names are, again,

1) Interoperable, providing a universal language that works across retailers, hospitals, states, and countries

2) Precise, as a generic name identifies a unique active ingredient

3). Systematic, offering patterns to help you teach those whom you care for

Let's learn the system I believe will revolutionize the way we think about medication names.

TABLE OF CONTENTS

Table of Contents

Table of Contents

INTRODUCTION

THE TOE-KNEE, T-O-E, K-N-E-E, PROBLEM

Imagine, for a minute, if the next time you went to the medical doctor, you *were* a licensed doctor. How would that appointment compare to previous appointments you've had? Would you feel and communicate better? This is a thought experiment, but it's possible to get closer to speaking like a doctor. What would be the first step?

Traditionally, doctors study organic chemistry and biochemistry before they learn pharmacology, or the study of drugs. But what if, instead of building drug names from organic and biochemistry parts, we built them from plain English words that represented images?

This book does just that so you can picture exactly how to pronounce and spell each medication name, and better remember them. Before I start with medication names, I want to outline a problem with personal names.

My name is Tony Guerra.

If I respell my first name as "toe, t-o-e", as in a digit from your foot, and "knee, k-n-e-e," as in the joint between your thigh and leg, you can pronounce both syllables. This builds a strong, memorable visual. There are, however, at least 13 different ways to spell the name Tony. How can I pronounce and spell Tony with a precise image?

I can say take the "t-o" from "toe," but what about the "n-y" sound? For Halloween, my three five-year-olds wore Rainbow Dash costumes, a popular cartoon pony. I could point to my daughters' outfits and say take the "n-y" from pony. (Note: the connection with "Tony" and "pony" made for a brutal middle school experience.) But, all kidding aside, it's an effective image, a toe and a pony. Tony.

My last name G-u-e-r-r-a, Guerra creates the same problem. While the double rolled "r" is something which is absent from English, I often ask people to pronounce my last name as Guerra, rhyming with Sarah. To create precision with spelling, however, I could say they should drop the "s-s-e" from the word "g-u-e-s-s-e-r, guesser" and add the r-a, ra from the word "aura, a-u-r-a". You can picture a winning guesser with an aura, aglow on the game show *Jeopardy* and with G-u-e-r-r-a handwritten in white on a blue electronic screen. Can this system work for medications?

A SYSTEM TO REPLACE RESPELLING

Right now, drug references use what's called respelling to help people pronounce drug names. For example, a reference might re-spell ibuprofen as e-y-e, eye, b-y-o-o, byoo, p-r-o-h, proh, and f-e-n, fen.

You can picture an eye, like an eyeball, but what is a b-y-o-o, byoo or a p-r-o-h, proh? A fen is a swamp, so that's a solid image. But, only literature nerds like me use fen. Instead, you can think of the first three letters of fender, as in a car bumper, if you prefer.

Unfortunately, respelling pulls you away from the correct spelling and creates two spellings, or orthographies if you want to get technical. Respellings change unpredictably across different websites and texts. My system works to prevent medication errors by introducing spelling medications as parts of English words which you already know, instead of using respellings that you don't.

Thus, I would write ibuprofen as "i" as in the letter "i;" b-u, bu as the first two letters in bu(reau), a chest of drawers. Then think p-r-o, pro is part of pro(ton), the positively charged atomic particle; and f-e-n, fen, as in swamp, or imagine a car fender sinking into a swamp.

Can you picture the letter "i" on a chest of drawers with a big plus sign for proton in a swamp? It's a silly image, but if I asked you to spell ibuprofen, could you do it now?

i
b-u
p-r-o
f-e-n

More importantly, can you now teach someone else to pronounce and spell ibuprofen so they remember both exactly?

You could say "i" as in the letter "i;"
b-u, bu, as the first two letters in bu(reau);
p-r-o, pro as in a pro(ton); and
f-e-n, fen, as in swamp.

This system is a learning *and* teaching tool. It also addresses the vague "schwa" sound.

THE PROBLEM WITH THE UNSTRESSED SCHWA

You might already know some parts of medication names. Calcium from calcium carbonate might be familiar as part of milk or bones. But how do you teach someone to spell and pronounce carbonate?

C-a-r from car, b-o from bo(ne), as in your shinbone, and n-a-t-e, nate, as part of the word (in)nate, or Nate, short for Nathan. This provides the correct spelling and pronunciation. But a dictionary may list the "o" in carbonate as an upside down "e."

That upside down "e" is a schwa, an unstressed syllable that often sounds like u-h, uh. The problem is all five vowels can form the schwa sound: "a" in comma, "e" in enemy, "i" in experiment, "o" in brother, and "u" in campus. The vowel between "b" and "n" in carbonate becomes ambiguous when transferring speech to text. By creating subtle pronunciation shifts to associate the "b-o" from bone, you attach the correct spelling to a concrete image. Knowing which vowel the schwa represents is vital in medication stems.

THE IMPORTANCE OF DRUG STEMS

In this book, I will use the terms prefix, suffix, and infix. Generic drug names are invented words and do not always conform to the rules of English. The United States Adopted Names Council properly calls each prefix and suffix that has meaning (e.g., cef- represents cephalosporins or –cillin represents penicillins) a stem.

In the case of penicillin, -cillin, c-i-l-l-i-n is the stem and peni-, p-e-n-i, is a prefix that differentiates peni*cillin* from other penicillin antibiotics such as amoxi*cillin* or ampi*cillin*, etc.

An infix is inside the word to make the classification more specific. The proper stem for a quinolone antibiotic is –oxacin, o-x-a-c-i-n, but cipro*floxacin* has the infix f-l, to classify it further as a *fluoro*quinolone, one that contains a fluorine atom.

Stems work as a heuristic, a shortcut to help us recognize a drug's class. If ten medications all end in o-l-o-l, –olol, we know these medications are likely beta-blockers.

The schwa, however, makes different spellings on paper have identical sounding endings.

For example, loratadine, l-o-r-a-t-a-d-i-n-e, and famotidine, f-a-m-o-t-i-d-i-n-e both end with the sound tuh-dine, but end with t-a-dine and t-i-dine respectively.

A drug that ends in t-i-d-i-n-e, tidine is an H_2-blocker to reduce acid. A drug that ends in a-t-a-d-i-n-e, atadine is an H_1-blocker, for allergies. Orthographically, or by spelling, we see the difference, but to have two drugs with identical pronunciation endings while in different drug classes is very bad. I recommend changing many dictionary pronunciations because I believe we've never taken time to ask, "What is the *safest* way to pronounce medications?"

I believe removing the schwa when possible is one way to make pronunciation safer. With generic drug names, I think it is right to be prescriptive. Car-bo-nate is clearer than car-buh-nate, as fa-mo-ti-dine lends less confusion than fa-mo-tuh-dine.

A DEADLY EXAMPLE

Many medication names have ambiguous spellings if based from oral pronunciation. The similar sounding hydromorphone and morphine can be deadly. The morphone, m-o-r-p-h-o-n-e from hydromorphone is only one letter different than morphine, m-o-r-p-h-i-n-e, but represents a ten times overdose that causes respiratory depression and has killed patients. A distracted health care provider might overlook this detail. However, by using images for each syllable, hydromorphone and morphine become visually different. With high risk, easily confused items, you can use this system to make labels for a medicine dispensing machine safer.

Hydromorphone

hy	= hy(brid), hy(phen), magnesium hydroxide
dro	= dro(ne), hydrocortisone
mor	= mor(e), Mor(se)(code), hydromorphone
phone	= phone, hydromorphone

Morphine

mor	= mor(e), Mor(se)(code), hydromorphone
phine	= (So)phi(a)(du)ne, morphine

How can we possibly learn so many medication names? Instead of repeating short phrases or words, like a language book, or the *repite, por favor* from Spanish class, I'm going to walk you through the story of building this system. Through narrative, you'll learn the language organically. We'll start with twenty-five children's medicines, just one shy of the number of letters in the alphabet. We'll use the sounds from these OTC drugs that most people have ready access to, much like an alphabet. Let me explain.

My daughters are five, five, and five. They sit in the same kindergarten class. They're learning sight words like a, an, and, be, by, the, then, them, etc. They're called sight words because children should know them on sight. Sometimes letters from one sight word form part of another. A-n, an is part of a-n-d, and. T-h-e, the is part of t-h-e-m, them. Generic drug names work similarly.

Each time we learn more medication names and their parts, we can build new names. Pharmacists handle many drugs on a daily basis, picking up parts of drug names and intuitively applying them to others. We'll do the same with what I call the Generic to English Translation System (GETS) I built. I like the acronym GETS because my goal is that everyone "gets" or understands generic naming and pronunciations after reading this book.

THE NUTS AND BOLTS OF THE GETS SYSTEM

This book will build a mental database of syllable parts by reviewing:
1. 25 common over-the-counter children's medications
2. 27 prescription children's drugs
3. 350 common medications
4. 20 children's oncology medications
5. 30 newer drugs on the market

The goal isn't to memorize the medicines. The goal is to learn how generic names are interoperable, precise, and systematic. You'll pick up quite a bit on the way.

With each medication, I will break the sounds apart and find a single visual word (if possible) or a combination of two words that makes each sound in the drug name. You won't have to do any rote repetition in this book. Rather, the book allows you to sit back and read or listen. The story will help you remember the sounds and names.

Here are three notes about the system before we start.

1. If the word matches the sound in pronunciation *and* spelling — as in c-a-r, car from carbonate or f-e-n, fen from ibuprofen — I will make no changes.

2. If part of the word is a sound I want you to say, I will put the part you don't pronounce in parentheses. For example, I can put the "d" from "card" in parentheses and tell you to sound out the c-a-r, car, only.

3. If I can't make the sound from one word, I may need to use two. For example, we pronounce the d-i-n-e, dine from famotidine as a combination of the d-i, di, from b-i-r-d-i-e, birdie and n-e, ne from t-u-n-e, tune to make "dine." This rhymes with d-e-a-n, dean, or b-e-a-n, bean.

MEDICATION ERRORS' HIDDEN PREVALENCE

Medication errors are the third leading cause of death in the United States behind heart disease and cancer. Many errors are preventable. A recent study reported that medication errors are a much larger problem than we originally thought, and intuitively this makes sense.

When someone passes away, the cause of death is a heart attack or seizure, or some pathophysiologic condition. To write "medication error caused heart attack" as the cause, for example, is stating blame. This accusation requires evidence a health professional might lack. Further, if they are wrong, or right without enough evidence, the discrepancy can land them in legal trouble. So, let's work to prevent errors altogether. Let's bring everyone on board — including the patient, and especially if they are a child.

BRINGING CHILDREN INTO THE HEALTH CARE TEAM

Let's finish this chapter with a story about my triplet daughter Teagan, and how she caught her first medication error. She was four-years-old.

"That's not my medicine," Teagan said.

"How do you know?" I asked.

"My medicine's purple."

Teagan smiled, smug and ecstatic that she'd caught me.

It's not that she doesn't like me—it's that she doesn't prefer me. When I come home from work, my other two triplet daughters, Brielle and Rianne, run and hug me. Teagan stands gymnast erect, crosses her arms, and shouts, "I love Mommy!"

My wife and I are pharmacists. Our daughter catching a medication error is a proud moment. It's like another parent's kid scoring a first soccer goal or strikeout or touchdown. However, I hadn't made a mistake. Like many households, we have a shelf full of medicine, and the pink acetaminophen, brand Tylenol, had just happened to come out before the purple fexofenadine, brand Allegra. But I wondered if the girls could pronounce the generic name.

In bringing children to the drug store, they often become an obstacle. They want attention, wonder why you are taking so long, and contribute to potential medication errors. Can parents engage their child as part of the process? We can at least be more precise than naming medicine by color.

The three girls are in a YouTube video where I ask them to pronounce acetaminophen.

"Can you say acetaminophen?" I ask, sounding the word out slowly and breaking it at the vowels: a ce ta mi no phen. Teagan shakes her head "no" before I start filming.

Introduction

I'm not sure if she's saying, "I can't pronounce it," or "I won't pronounce it—for you—but I would for Mom."

They giggle, give it a shot, and at 29 seconds, Brielle gets it. How?

They don't care about mistakes while learning. They laugh the mistakes off, work with each other, and continue until they succeed.

Most adults, however, don't or won't try. I understand. Who wants to sound dumb or make a mistake? However, I've found, as with singing in church, I'll do many things that would otherwise embarrass me when I'm modeling behavior for my kids. Even if you are reluctant, I've created a way that you can compare your recording against my online pronunciation later in the book. You can do this privately at home.

Chapter A - 25 Over-the-counter (OTC) children's medications

Making sure two people check every dose

A prior conversation with my pastor helped me decide which drugs to include in this first section. I have three children. He has six. We talked about how parents of multiple children divide tasks with their spouses. When medicine meets exhausted parents, however, you want a second set of eyes.

But what if you're a single parent, or your spouse works where you can't get a hold of them? Maybe they are military and deployed, working somewhere loud, or unavailable for any other number of reasons. My grandparents were U.S. Navy. I can only imagine the relationship difference if they'd had the technology to communicate about their children, my mother and uncle, while separated. The solution is that you still need to find someone to serve as a second pair of eyes.

So, know this: I worked as an overnight pharmacist and I loved getting middle of the night questions and talking to a human instead of myself. Don't feel like you are bothering the pharmacist.

Why are children's over-the-counter sections so confusing?

In the U.S., when doctors write prescriptions, pharmacists check them. In a hospital, nurses check meds before giving them out. In the United States, we have the freedom to buy many over-the-counter children's medicines without any barrier. If a parent, often exhausted or distracted, walks into a pharmacy and puts the medicine in their cart, though, no one checks it for accuracy — a cashier simply bags it.

Most drugstore and grocery chains rely on plans that floor staff use to tell them where to put medicines. Sometimes the drug's shelf location doesn't align with its signage. I went home to North Potomac, Maryland and, in the pharmacy, I saw the "antidiarrheal" sign above a laxative, and an "analgesics" sign above antacids. You can't rely on signage. Depend on the active ingredient list on the back of medication boxes. That's what I want to teach you to read.

Let's pull up two stools and sit down in front of the pediatric section in our local pharmacy, just as I would with a student. I'm going to have to also bring along my cheater glasses, because I can't read most medication labels anymore. I'm 44 years old and have presbyopia, a worsening of my near vision. I have to take pictures of medication labels with my iPhone, then expand them to read the tiny font. It's worth the trouble. Small font does not mean small in importance for generic names.

MEDICATIONS

Antacids

1. Calcium Carbonate (Children's Pepto)

cal'	= cal(orie), (lo)cal(e), calcium carbonate
ci	= (pronun)ci(ation), calcium carbonate
um	= (g)um, calcium carbonate
car'	= car, car(d), calcium carbonate
bo	= bo(ne), calcium, carbonate
nate	= (in)nate, Nate, calcium carbonate

We begin with 1) **calcium carbonate,** a generic antacid. Calcium carbonate relieves acid reflux, increasing the pH and decreasing the acidity that can cause heartburn. But, if we want a clerk to point us to the medication, it's easier to ask for the shorter brand name, such as **Children's Pepto.** Careful, though—the adult calcium carbonate's brand name is **Tums.** Children's Pepto may be under a white signpost that reads "digestive," in a separate children's section, or in some other section with gastrointestinal medicines.

The goal is to preserve a medication's correct spelling so that, when you picture the drug name in your mind's eye, you spell it correctly.

Pronounce the c-a-l, cal from calcium carbonate as the first syllable in calorie, as in "I'm trying to cut calories; " c-i, ci in calcium like the middle of the word pronun-ci-ation; and the u-m, um, comes from gum. Calorie, pronunciation, gum. Cal ci um. Calcium.

Easy, right? Well, what if it wasn't. Is there a backup plan? Yes. It's called **backbuilding**. If you struggle to pronounce something going forward in English, often going backwards helps you get the rhythym right. Backbuilding means you say the last syllable first. Then you say the last and the second to last. Then say the last, second to last, and third to last until you've completed the words. It works with medical terminology too, if you've had trouble with those Latin and Greek terms.

Calcium becomes: Um. Ci um. Cal ci um. Calcium. For carbonate, use c-a-r, car, an automobile; the b-o, bo from bone, as in a shinbone; and n-a-t-e, nate from innate. Car, bone, innate. Nate. Bo nate. Car bo nate. Carbonate. Calcium carbonate.

Constipation – Saline laxative, stool softener, and osmotics

2. Magnesium Hydroxide (Pedia-Lax chewable tablets saline laxative)

mag	= mag(azine), magnesium hydroxide
ne'	= (mo)ne(y), k(ne)e, magnesium hydroxide
si	= (enthu)si(astic), (tran)si(ent), magnesium hydroxide
um	= (g)um, calcium carbonate
hy	= hy(brid), hy(phen), magnesium hydroxide
drox'	= dr(ink) (b)ox, dr(op) (b)ox, magnesium hydroxide
ide	= (sl)ide, magnesium hydroxide

Like the Pepto described in the introduction, 2) magnesium hydroxide and 3) docusate sodium are subject to "brand extensions."

Pedia-Lax chewable tablets saline laxative has, as its active ingredient, **magnesium hydroxide**, while **Pedia-Lax liquid stool softener** has as its active ingredient oral **docusate sodium**. Both relieve constipation.

The m-a-g, mag from magnesium hydroxide includes the first three letters in magazine; the n-e, ne comes from the middle of money; s-I, si from enthusiastic, and u-m, um, comes from gum. Picture opening the magazine *Money* enthusiastically and then finding a wet wad of gum inside. Magazine, money, enthusiastic, gum. Um. Si um. Ne si um. Mag ne si um. Magnesium.

We build hydroxide from the h-y, hy in hybrid car; the d-r from drink and the o-x from box to make "drox;" and the i-d-e, ide from slide. Picture a hybrid's wheels turning as you spill a drink box and slide on an icy road. Hybrid, drink box, slide. Ide. Drox ide. Hy drox ide. Magnesium hydroxide.

Constipation – Stool softeners

3. Docusate sodium (Pedia-Lax liquid stool softener)

do'	= do(t), do(dge), docusate sodium
cu	= cu(be), cu(re), cu(te), docusate sodium
sate	= (pul)sate, sate, (compen)sate, docusate sodium
so'	= so(fa), so, docusate sodium
di	= (bir)di(e), (per)di(em), docusate sodium
um	= (g)um, calcium carbonate

Stool softeners like **docusate sodium** bring in water to make a stool easier to pass. The d-o, do from dot, as in "dot your i's"; the c-u, cu from cube, as in ice cube; and the s-a-t-e, sate from pulsate, as in "the dance floor would pulsate." Dot, cube, pulsate. Sate. Cu sate. Do cu sate. Docusate.

Sometimes you'll see a salt form, like sodium, s-o-d-i-u-m, after a drug name. Salt forms help with drug absorption or impart properties like short versus long-acting. For example, some blood pressure and sleeping pills last for 6 hours or 12 hours based on the salt form.

While sodium is an English word, it's readily broken into sound parts by taking: the s-o, so, from sofa; the d-i, di, from birdie; and the u-m, um, from gum. Sofa, birdie, gum. Um. Di um. So di um. Sodium. Docusate sodium.

Strangely, one day our daughters had a sofa, a live birdie, and gum all in our living room at once — the bird escaped unharmed.

Constipation – Osmotic

4. Polyethylene glycol (PEG) 3350 (MiraLax)

po	= po(t), po(lygon), polyethylene glycol
ly	= (li)ly, (po)ly(gon), polyethylene glycol
eth'	= (m)eth(od), polyethylene glycol
yl	= (s)yl(lable), polyethylene glycol
ene	= (sc)ene, polyethylene glycol
gly'	= gly(cemic), polyethylene glycol
col	= col(lar), Col(orado), polyethylene glycol

While **polyethylene glycol** as **MiraLax** provides gentle relief and easy access as an over-the-counter medication, make sure you've had the doctor provide you with the correct dosage. Generally, we divide dosages into groups for patients from 6 months to 1 year, then 1 to 3 years, 4 to 7 years, and 8 years and up. We adjust based on age, size, constipation severity, and diet. This includes water intake relative to the types of stools the patient produces.

Polyethylene glycol comes in white bottles with dark purple lettering. The colored screw-on top serves as a measuring cup for the powder inside. The brand name alludes to miracle plus laxative. MiraLax.

The temptation with polyethylene glycol is to use the p-o-l-y, poly, which means "many." A polygon is a many-sided geometric shape. But shapes with an undefined number of sides are tough to picture. It's easier to use the p-o, po, from pot and the l-i, li, from lily, and picture a pot with a lily in it. The e-t-h, eth, is in the word method; the y-l, yl, sits in syllable; and the e-n-e, ene, ends scene.

Picture an actor holding a flower pot with a lily. He says his method is to plod through each syllable to remember his role in a scene. Pot, lily, method, syllable, scene. Ene. Yl ene. Eth yl ene. Ly eth yl ene. Po ly eth yl ene. Polyethylene.

A glycol is another chemical compound, but I think of the g-l-y, gly, from the glycemic index, where the number matches to how a food affects your blood glucose or blood sugar. For c-o-l, col, I think of a business shirt's neck collar with sugar on it. Glycemic, collar. Col. Gly col. Polyethylene glycol.

Antidiarrheals

5. Loperamide (Imodium A-D Children's Antidiarrheal Liquid)

lo	= (Co)lo(rado), (s)lo(w), loperamide
pe'	= pe(n), pe(ril), pe(t), loperamide
ra	= (au)ra, (cob)ra, (Op)ra(h), loperamide
mide	= (gu)m(sl)ide, loperamide

The bright green bottle with two kids' heads on it holds the mint flavored liquid **loperamide,** brand **Imodium A-D, Children's Antidiarrheal Liquid**. If you can think of loperamide as low-peristalsis, you can remember that it slows bowel movement. You can also relate the brand **Imodium** to the word immobile (to stop things from moving), as well.

Loperamide takes the l-o, lo, from Colorado; the p-e, pe, from pen; the r-a, ra, from aura; and mide, which is not in any English word I could find. The last two syllables — r-a, ra and m-i-d-e, mide — deserve commentary.

The schwa, or unstressed syllable, is an obstacle in spelling medicines correctly. The u-h, uh, sound is an interjection like u-m, um, or a-h, ah, and so on. In a-u-r-a, aura the last "a" is this uh sound. The problem is that all the vowels (a, e, i, o, and u) can make the u-h, uh sound, so someone hearing the name has no way of knowing which letter the schwa represents. In this case we use aura because we know aura's spelling of a-u-r-a.

To get m-i-d-e, mide, I had to combine two words — taking the "m" from gum and the "ide" from slide, picturing sticky gum at the end of a slide.

In Colorado's Rocky Mountain National Park, the ranger traced our hike through Glass Lake to Sky Pond with a neon pink pen. We had the beginner aura, happy to let gum slide off our shoes and get hiking. With a long hike and no rest rooms, loperamide is a good idea. Colorado, pen, aura, gum, slide. Mide. Ra mide. Pe ra mide. Lo pe ra mide. Loperamide.

Antigas

6. Simethicone (Mylicon)

si	= si(t), simethicone
meth'	= meth(od), simethicone
i	= i(t), simethicone
cone	= cone, (pine)cone, simethicone

Dairy for a lactose intolerant person eating an ice cream cone might call for an antigas medicine like **simethicone,** brand **Mylicon.** S-i, si, from sit and not stand, meth from method, and i, from i-t, it, and cone is a word like an ice cream cone. Sit, method, it, cone. Cone. I cone. Meth i cone. Si meth i cone. Simethicone.

Note: in this book, I'll tell you to break syllables at the vowels. One exception includes m-e-t-h, meth in the simethicone ending with consonants.

OTC Analgesics – Non-narcotic

7. Acetaminophen [APAP] (Infant's Tylenol, Children's Tylenol)

a	= (comm)a, (aur)a, acetaminophen
ce	= ce(iling), acetaminophen
ta	= (da)ta, (sona)ta, (quo)ta, acetaminophen
mi'	= mi(nt), mi(lk), acetaminophen
no	= (can)no(n), (ca)no(n), acetaminophen
phen	= (hy)phen, acetaminophen

Make sure that, with analgesics and NSAIDs, you get the infant's or children's formulation that matches your child's age. Always double-check with a pharmacist.

Acetaminophen is a non-narcotic analgesic, contrasting with medications like morphine that are narcotic analgesics. Just as ibuprofen is a non-steroidal, acetaminophen is a non-narcotic. It's like defining nonfiction by saying it's not fiction.

Acetaminophen's bright red box with white lettering is easy to spot on pharmacy shelves. The generic name, acetaminophen, and the brand name, **Tylenol**, come from parts of the chemical name, N-**acetyl**-para-**amino**-**phenol**.

When I started to write this second book, the first time I had a feeling this system would work came when I asked my daughters to say acetaminophen — as I talked about earlier. As now five-year-olds, making most of the generic name syllables consonant-vowel seems to help pronunciation in general.

For acetaminophen, take the "a" from comma; the c-e, ce from ceiling; t-a, ta from data; m-i, mi from mint; n-o, no from cannon ball; and p-h-e-n, phen from hyphen. Comma, ceiling, data, mint, cannon, hyphen. Phen. No phen. Mi no phen. Ta mi no phen. Ce ta mi no phen. A ce ta mi no phen. Acetaminophen.

pa	= pa(ir), pa(re), enoxaparin
ra	= (au)ra, (cob)ra, (Op)ra(h), loperamide
ce'	= ce(iling), acetaminophen

ta	= (da)ta, (sona)ta, (quo)ta, acetaminophen
mol	= mol(lusk), (enty)mol(ogy), methocarbamol

Acetaminophen is called paracetamol in many countries outside the United States. With both acetaminophen and paracetamol, we pronounce the two letter "a's" as "uh." We can just as easily find common English words for paracetamol as acetaminophen. Take the p-a, pa from pair of socks; the r-a, ra from aura; the c-e, ce from ceiling; the t-a, ta from data; and the m-o-l from mollusk. Pair, aura, ceiling, data, mollusk. Mol. Ta mol. Ce ta mol. Ra ce ta mol. Pa ra ce ta mol. Paracetamol.

OTC Analgesics – NSAIDs

8. Ibuprofen (Infant's Advil, Children's Advil, Infant's Motrin, Children's Motrin)

i	= "i", ibuprofen
bu	= bu(reau), ibuprofen
pro'	= pro(ton), pro(be), pro(active), ibuprofen
fen	= fen, fen(der), ibuprofen

Ibuprofen is a nonsteroidal anti-inflammatory drug, NSAID, like **aspirin**, but ibuprofen is safe for children and aspirin is not. If you see a patient scouring the adult ibuprofen section for children's ibuprofen, they may not realize it's in a different section than the adult medications. And within the children's section, we often distinguish children's and infant's formulas. Asking for a pharmacist's help is proper.

Pronounce "i" as the letter "i", and b-u, bu as in bureau. Bureaus are like desks. Then p-r-o, pro as in proton and f-e-n, fen, a swamp. Fen. Pro fen. Bu pro fen. I bu pro fen. Ibuprofen.

Next, we move onto first generation antihistamines. Again, these are the antihistamines that do make you sleepy.

Antihistamines – 1st-generation (brompheniramine, chlorpheniramine, diphenhydramine)

In a comedy about a dating coach named Hitch played by Will Smith, there was a scene where he went to the pharmacy's over-the-counter aisle to treat his food allergies. He looked into the security mirror, saw his severely swollen face, and then knocked over half of the medicines on the shelf before drinking a good helping of pink diphenhydramine, brand Benadryl.

Unfortunately, he likely took the wrong medicine, but it was good comedy. An oral liquid works more slowly than an intramuscular injection. A person having a rapid allergic reaction would likely need an epinephrine injection. The story reminds us to seek help even in this self-service department.

Brompheniramine, chlorpheniramine, and diphenhydramine are all first-generation antihistamines. They act on histamine-1 receptors to reduce common allergy symptoms like sneezing, itching, and hives.

A first-generation drug is the first in a class to show a desired activity. Later generations form a successive line of derivatives from the original drugs, often with fewer side effects due to molecular differences.

9. Brompheniramine (Dimetapp Cold and Cough, etc.)

brom	= brom(ine), brompheniramine
phe	= phe(nomenon), brompheniramine
ni'	= ni(ght), brompheniramine
ra	= (au)ra, (cob)ra , (Op)ra(h), loperamide
mine	= (hista)mine, brompheniramine

Brompheniramine is in many combination cold medications and has four new syllables. It's b-r-o-m, brom from bromine of the Periodic Table of Elements. Then, p-h-e, phe from phenomenon, an unusual event like a large full moon.The n-i, ni comes from night and the r-a, ra from aura with the schwa sound.

Now, m-i-n-e, mine we pronounce like that bully is mean, but spelled m-i-n-e, as if you will dig a tunnel. How do we remember it? Histamine has m-i-n-e, mine pronounced as "mean" also. Histamine causes allergy symptoms and that's mean. Bromine, phenomenon, night, aura, histamine. Mine. Ra mine. Ni ra mine. Phe ni ra mine. Brom phe ni ra mine. Brompheniramine.

10. Chlorpheniramine (Dimetapp Long-Acting Cough Plus Cold, etc.)

chlor	= chlor(ine), chlorpheniramine
phe	= phe(nomenon), brompheniramine
ni'	= ni(ght), brompheniramine
ra	= (au)ra, (cob)ra , (Op)ra(h), loperamide
mine	= (hista)mine, brompheniramine

Chlorpheniramine begins with c-h-l-o-r, chlor as in chlorine from a swimming pool. And the remainder is identical to brompheniramine, the "phe" from phenomenon, "ni" from night, "ra" in aura, and "mine" from histamine. Chlorine, phenomenon, night, aura, histamine. Mine. Ra mine. Ni ra mine. Phe ni ra mine. Chlor phe ni ra mine. Chlorpheniramine.

11. Diphenhydramine (Children's Benadryl Allergy)

di	= di(ce), diphenhydramine
phen	= (hy)phen, acetaminophen
hy'	= hy(brid), hy(phen), magnesium hydroxide
dra	= (hy)dra, diphenhydramine
mine	= (hista)mine, brompheniramine

We build **diphenhydramine** from d-i, di in dice, as in those for board games; p-h-e-n, phen from hyphen; then the h-y, hy, from hybrid; d-r-a, dra from hydra, the mythical many-headed monster; and m-i-n-e, mine from histamine. Dice, hyphen, hybrid, hydra, histamine. Di phen hy dra mine. Mine. Dra mine. Hy dra mine. Phen hy dra mine. Diphenhydramine.

OTC Antihistamines – 2nd-generation (Cetirizine, fexofenadine, loratadine)

Second generation antihistamine means the antihistamine is most likely *not* going to make a person become drowsy. The first and older generation usually causes sedation.

12. Cetirizine (Children's Zyrtec Allergy)

ce	= ce(nt), ce(nter), cetirizine
ti'	= ti(c), (plas)ti(c), cetirizine
ri	= ri(nse), (Flo)ri(da), cetirizine
zine	= (maga)zine, cetirizine

Cast in a bright lime green box, this *second*-generation antihistamine also acts on histamine-1 receptors to reduce allergy symptoms. Unlike diphenhydramine, **cetirizine** produces significantly less drowsiness. Recognize that cetirizine may take several days to provide symptomatic improvement, though, whereas diphenhydramine begins working within a few hours.

The c-e, ce comes from the word cent and the t-i, ti is like a tic from a movement disorder like Parkinson's disease. The r-i, ri comes from rinse or Florida. Z-i-n-e, zine is the last sound in magazine. Cent, tic, rinse, magazine. Zine. Ri zine. Ti ri zine. Ce ti ri zine. Cetirizine.

Note, a dictionary might list the second "i" in cetirizine as the schwa sound, but doing that creates spelling ambiguity.

13. Fexofenadine (Children's Allegra)

fex	= (Ponti)fex, fexofenadine
o	= "o," fexofenadine
fe'	= fe(n), fexofenadine
na	= (bana)na, fexofenadine
dine	= (bir)di(e)(tu)ne, fexofenadine

Fexofenadine starts with fex, which is part of the word pontifex, an ancient Roman priest. The "o" is the sound the letter "o" makes. F-e, fe comes from fen, a swamp, but we're going to drop the "n" from fen because we want to keep vowels at the end of syllables.

How to Pronounce Drug Names

The n-a, na includes the schwa sound, as in the end of banana. The last sound is d-i-n-e, dine, spelled like dine, as in to eat, and pronounced like dean, as in the head of a college. To solve this, I've taken the d-i, di from birdie and the n-e, ne from tune to make "dine". Pontifex, o, fen, banana, birdie, tune. Dine. Na dine. Fe na dine. O fe na dine. Fex o fe na dine. Fexofenadine.

Why don't we just use fen-a-dine instead of fe-na-dine? We don't want to leave a letter like "a" alone with the schwa sound if we don't have to and then have a five-letter a,e,i,o, or u multiple-choice when trying to spell a medicine.

14. Lor<u>atadine</u> (Children's Claritin)

lo	= (Co)lo(rado), (s)lo(w), loperamide
ra'	= ra(ttle), loratadine
ta	= (da)ta, (sona)ta, (quo)ta, acetaminophen
dine	= (bir)di(e)(tu)ne, fexofenadine

Loratadine has a recognized stem in its generic name, "-atadine, a-t-a-d-i-n-e", and its brand name is **Claritin**. The catchy marketing slogan, "Get Claritin Clear," helps students remember the drug clears up allergies. Claritin's packaging displays a green field and a *clear* blue sky. This subtly reminds patients that with Claritin, seasonal allergies don't have to stop them from enjoying the outdoors.

In loratadine, the l-o, lo comes from the state of Colorado. This r-a is not the "ra" from aura, though; it's r-a, ra from from a baby's rattle. Then t-a, ta from data and the d-i-n-e, dine we previously assembled from birdie and tune. Colorado, rattle, data, birdie, tune. Dine. Ta dine. Ra ta dine. Lo ra ta dine. Loratadine.

You will hear some patients say "lorata-dine," as in to eat. What are you going to do to correct them? Pronounce it correctly when reflecting their question, but be respectful and subtle.

Allergic rhinitis nasal steroids (fluticasone, triamcinolone)

Let's start with pathophysiology. What is allergic rhinitis? How do we treat it?

22

Allergic rhinitis is an inflammation (-*itis*) of the nose (*rhin*-). We treat it with a local nasal steroid like **fluticasone** or **triamcinolone**. Nasal steroids don't work right away like a topical decongestant; rather, it takes weeks before a patient feels relief.

15. Fluticasone (Children's Flonase)

flu	= flu(id), fluticasone
ti'	= ti(c), (plas)ti(c), cetirizine
ca	= (or)ca, fluticasone
sone	= so(fa), (du)ne, fluticasone

Fluticasone begins with f-l-u, flu from fluid, and goes on to t-i, ti from tic, a movement disorder, and c-a, ca with the schwa that comes at the end of orca, a whale.

The s-o-n-e, sone sounds like s-e-w-n. To resolve this spelling issue, I used the s-o, so from sofa and the n-e, ne from dune to build "sone." Fluid, tic, orca, sofa, dune. Sone. Ca sone. Ti ca sone. Flu ti ca sone. Fluticasone.

16. Triamcinolone (Children's Nasacort Allergy 24HR)

tri	= tri(angle), triamcinolone
am	= (l)am(p), triamcinolone
ci'	= (s)ci(ntillate), triamcinolone
no	= (can)no(n), (ca)no(n), acetaminophen
lone	= lone, (a)lone, (c)lone, triamcinolone

The brand name **Nasacort** comes from nasa for nose plus cort, part of corticosteroid, to make an anti-inflammatory for your nose.

In **triamcinolone**, the t-r-i, tri is from triangle, a three-sided geometric shape; the a-m, am comes from lamp, a light source; and the c-i, ci comes from scintillate, something very exciting. The n-o, no comes from cannon ball with the schwa, and l-o-n-e, lone from lone wolf. Triangle, lamp, scintillate, cannon, lone. Tri am ci no lone. Triamcinolone, a steroid for allergic rhinitis.

Topical steroids

17. Hydrocortisone (Children's Cortizone-10)

hy	= hy(brid), hy(phen), magnesium hydroxide
dro	= dro(ne), hydrocortisone
cor'	= cor(n), cor(k), hydrocortisone
ti	= ti(c), (plas)ti(c), cetirizine
sone	= so(fa), (du)ne, fluticasone

One topical steroid for children is **hydrocortisone**. Start with h-y, hy from hybrid and d-r-o, dro from drone, the flying machine that can film a football game with its camera. Take the c-o-r, cor from Iowa sweet corn; the t-i, ti from tic, a movement disorder; and the s-o-n-e, sone from sofa and dune. Hybrid, drone, corn, tic, sofa, dune. Sone. Ti sone. Cor ti sone. Dro cor ti sone. Hy dro cor ti sone. Hydrocortisone.

OTC Antitussives / Expectorants

18-19. Guaifenesin/dextromethorphan (Children's Mucinex Cough)

dex	= dex(terity), dextromethorphan
tro	= (me)tro, (re)tro, dextromethorphan
meth	= meth(od), simethicone
or'	= or, (st)or(e), dextromethorphan
phan	= (or)phan, dextromethorphan

Dextromethorphan with **guaifenesin** is an anti-cough / expectorant combination. Dextromethorphan starts with d-e-x, dex, from dexterity. *Dexter* is Latin for right. T-r-o, tro is part of metro, as in the Washington D.C. metrorail; m-e-t-h, meth comes from method; o-r, or is itself a word, a conjunction like "and;" and p-h-a-n, phan comes from orphan, a child without parents. Dexterity, metro, method, or, orphan. Phan. Or phan. Meth or phan. Tro meth or phan. Dex tro meth or phan. Dextromethorphan.

guai	= gua(va)"i", gua(camole)"i", guaifenesin
fe'	= fe(n), fexofenadine
ne	= ne(t), guaifenesin
sin	= (ba)sin, sin, guaifenesin

Guaifenesin's g-u-a-i, guai isn't found in English, but the g-u-a, gua is in guava plant or guacamole for tortilla chips. Add the letter "i" to make "gua-i." Take the f-e, fe from fen; then n-e, ne from net, like a soccer net; and s-i-n, sin from basin, a dried up lake or a place in the kitchen to store water. Guava + i, fen, net, basin. Sin. Ne sin. Fe ne sin. Guai fe ne sin. Guaifenesin.

Note: a dictionary might list the second "e" in guaifenesin as a schwa sound rather than an "e."

Decongestants

20. Phenylephrine (PediaCare Children's Decongestant)

phen	= (hy)phen, acetaminophen
yl	= (s)yl(lable), polyethylene glycol
eph'	= (z)eph(yr), phenylephrine
rine	= (doct)rine, (u)rine, phenylephrine

Phenylephrine is an intranasal product that relieves sinus congestion.

Phenylephrine begins with the p-h-e-n, phen from hyphen; moves to y-l, yl, from syllable; e-p-h, eph from zephyr, a gentle wind; and r-i-n-e, rine from doctrine, a mandate or law. The last r-i-n-e is in urine also, but my girls would say, "urine and nasal decongestant — gross." Hyphen, syllable, zephyr, doctrine. Rine. Eph rine. Nyl eph rine. Phe nyl eph rine. Phenylephrine.

Topical antibiotics

21-23. Bacitracin / Neomycin sulfate / Polymyxin B (Neosporin)

ba	= ba(t), bacitracin
ci	= (s)ci(ntillate), triamcinolone
tra'	= tra(y), bacitracin
cin	= cin(der), (s)cin(tillate), bacitracin
ne	= (mo)ne(y), k(ne)e, magnesium hydroxide
o	= "o," fexofenadine
my'	= my, my(osin), neomycin sulfate

cin	= cin(der), (s)cin(tillate), bacitracin
sul'	= sul(len), neomycin sulfate
fate	= fate, neomycin sulfate
po	= po(t), po(lygon), polyethylene glycol
ly	= (li)ly, (po)ly(gon), polyethylene glycol
myx'	= myx(edema), polymyxin B
in	= in, (t)in, polymyxin B
B	= "B"

Above boxes of superhero and princess Band-Aids is a brightly colored golden box labeled **Neosporin**. If I listed the generic components—**neomycin, polymyxin B**, and **bacitracin**—you might not recognize them right away. But, by taking pieces of each generic antibiotic name, you can form the brand **Neosporin**. First, take "n-e-o-s" from **neomycin sulfate**, then the "p-o" from **polymyxin**, and finally the "r-i-n" from **bacitracin**.

Neosporin ointment covers minor cuts and scrapes to help prevent infection. Neosporin has three active ingredients: bacitracin, neomycin sulfate, and polymyxin B.

For **bacitracin**, take the b-a, ba from bat, that which lives in a cave, and c-i, ci from from scintillate, and triamcinolone, the nasal steroid. The t-r-a, tra is the first three letters of t-r-a-y, tray; then c-i-n, cin from cinder by the fireplace or the first three letters of the sister from *Grimm's Fairy Tales*. Bat, scintillate, tray, cinder. Cin. Tra cin. Ci tra cin. Ba ci tra cin. Bacitracin, the first antibiotic in the triple combination.

The second is **neomycin** with n-e, ne from money; "o" is the letter "o;" m-y, my is a word itself; and then c-i-n, cin from cinder. Money, o, my, cinder. Cin. My cin. O my cin. Ne o my cin. Neomycin.

Careful, it's neomy-cin, not neomy-o-cin, a common error that comes from difficulty getting from the "y" to the "c" without inserting a vowel.

Sulfate comes from s-u-l, sul from sullen, a sad state, and f-a-t-e, fate, something you might see in a crystal ball, your fate.

Polymyxin B comes from p-o, po from cooking pot; l-y, ly from lily pad; m-y-x, myx from myxedema that sounds like cookie m-i-x, mix; and then i-n, in, and the capital letter "B." Pot, lily, myxedema, in, B. B. In B. Myx in B. Ly myx in B. Po ly myx in B. Polymyxin B.

Note: Neosporin Plus Pain Relief has a picture of children on the box. This combination has a different active ingredient. The original Neosporin works for children over 2 years old in the US.

24. Permethrin (RID, Nix)

per	= per(son), permethrin
meth'	= meth(od), simethicone
rin	= rin(se), permethrin

Permethrin is classified as anti-lice and starts with the p-e-r, per from person, a human, and then m-e-t-h, meth from method, and r-i-n, rin from rinsing your hair. Person, method, rinse. Rin. Meth rin. Per meth rin. Permethrin.

25. Benzocaine (Baby Orajel)

ben'	= ben(d), benzocaine
zo	= zo(ne), (o)zo(ne), (gon)zo, benzocaine
caine	= c(at)(migr)aine, benzocaine

Note: An FDA Warning about benzocaine and methemoglobinemia exists. This book is about pronunciation and spelling, not therapeutic decision making. Please speak to your pharmacist before grabbing benzocaine and using it with a child under 2.

Our last drug name in the over-the-counter children's medication section is **benzocaine.** You could use the person's name, Benjamin, and shorten it to "Ben" to get the first three letters of this medication. But, I prefer to use the word b-e-n-d, bend and drop the "d."

The z-o, zo comes from end zone and c-a-i-n-e, caine comes from a combination of "c" from cat, a feline, and a-i-n-e, aine from migraine.

Cocaine was the first local anesthetic with the aine ending. But, I wanted a strong visual—a cat migraine. Bend, zone, cat, migraine. Caine. Zo caine. Ben zo caine. Benzocaine.

Chapter B - Prescription children's medications

In this chapter, I am going to review prescription children's medications. I chose drugs my own children have been on because the issue with calling something a children's medication is that many are used off-label, only about 25% of children's medicines get tested in children. That is, a physician prescribes it even though there have been no formal pediatric studies done, but practice evidence shows the therapy is effective. It's very tough to do tests on children — there are ethics involved, hence trouble getting good data.

 I have three five-year-old children who spent two months in the neonatal intensive care unit (NICU), so getting to 25 prescription medications they have been on was not hard. I will go through the pronunciations as I did in Chapter A, but now I'll focus on practical pediatric cases.

Chapter A was about keeping your kids safe when you get medications for them. This chapter is about becoming a knowledgeable parent when interacting with prescribers. Just as you want to turn every box around every time with over-the-counter drugs, your interactions with providers should be two-way conversations.

The triplets and spelling, pronunciation and mnemonics

Brielle, Rianne, and Teagan have distinct learning personas. Brielle, whose name fittingly rhymes with spell, will ask you to spell with her if she doesn't understand something. After we've read at night, she'll go on to do extra words on her iPad that are not on our current sight word list. She finds great joy in learning.

Rianne is adamant about correct pronunciation. She's wanted to correct my pronunciation of the planet J-a-k-k-u, Jakku from *Star Wars: The Force Awakens*, as from Ja-ku to Jak-ku. In English, if we have a word like *connect*, we only pronounce one of the n's, but she notices things like that. I pronounced Disney's Anna, pronounced like I'm "on" a snowdrift versus A-n-n, Ann-uh. I asked her to pronounce b-u-t, but, and she said: "With one t, that's okay, but if you use two t's, that's inappropriate; you should instead say 'bottom'."

Brielle has the spelling, Rianne the pronunciation, and Teagan, T-e-a-g-a-n, is about adding clever images and motion. Teagan is the joker and innovator of the group. She spits out memorable mnemonics and shares what immediately comes to mind.

Let's start with the first group of medicines and Teagan's stories.

TEAGAN – TROUBLE BREATHING

Beta-2 receptor agonist short acting (bronchodilator) and an anticholinergic bronchodilator

1-2. Albuterol / Ipratropium (DuoNeb)

al	= (s)al(iva), Al(exander), albuterol
bu'	= bu(reau), ibuprofen
ter	= ter(se), albuterol
ol	= (aw)ol, (alcoh)ol, albuterol
ip	= (s)ip, ipratropium
ra	= (au)ra, (cob)ra , (Op)ra(h), loperamide
tro'	= (me)tro, (re)tro, dextromethorphan
pi	= pi(ece), ipratropium
um	= (g)um, calcium carbonate

A beta-2 receptor agonist is a short-acting bronchodilator. The anticholinergic **ipratropium** is also a bronchodilator. **Albuterol** is the beta-2 receptor agonist, and that means there is a beta receptor (Beta, the second letter in the Greek alphabet); when *activated*, it signals bronchioles to open in the lungs.

Ipratropium bronchodilates work by *blocking* the cholinergic receptor. Why use two medications to open the lungs? With two medicines that both bronchodilate, you can lower doses to decrease toxicity.

Teagan and Brielle both had supplemental oxygen after their NICU stay. We had two oxygen generating machines and multiple green oxygen tanks to travel. When they struggled to breathe, giving them albuterol with ipratropium helped. The brand name DuoNeb comes from the nebulized medicine duo.

Let's review the pronunciation: the a-l, al in albuterol is from saliva or my middle name Alexander; the b-u, bu comes from bureau, and ibuprofen. The t-e-r, ter comes from terse. If your spouse is mad, she might speak tersely.

The ending of the military acronym AWOL, absent without leave, is the last syllable in albuterol, but you can use the o-l, ol from alcohol. However, I wanted stay away from alcohol because it's a chemical.

Saliva, bureau, terse, AWOL. Ol. Ter ol. Bu ter ol. Al bu ter ol. Albuterol.

The ipratropium takes i-p, ip from sip a drink and the r-a, ra comes from aura. The t-r-o, tro comes from metro rail in Washington D.C. and the p-i, pi sounds like puzzle piece. The u-m, um stems from gum. Sip, aura, metro, piece, and gum. Um. Pi um. Tro pi um. Ra tro pi um. Ip ra tro pi um. Ipratropium.

Leukotriene receptor antagonist

3. Monte<u>lukast</u> (Singulair)

mon	= mon(key), montelukast
te	= te(nt), motelukast

lu'	= lu(cid), montelukast
kast	= (Out)kast, (Di)kast, montelukast

Teagan needed additional help because she had pediatric apnea. A leukotriene receptor antagonist also helps asthmatics and other respiratory conditions. **Montelukast** doesn't have any syllables we've heard before and this is common with a new generic drug class. Let's look at montelukast, brand **Singulair**. The m-o-n, mon is from monkey; the t-e, te is from tent; the l-u, lu is from lucid, which means that things are clear in your brain; and then k-a-s-t, kast I took from the musical group O-u-t-k-a-s-t or an Athenian judge, a D-i-k-a-s-t. Monkey tent, lucid, Outkast. Kast. Lu kast. Te lu kast. Mon te lu kast. Montelukast.

Oral liquid steroid

4. Prednisolone (Orapred)

pred'	= pred(ator), pred(ict), prednisonolone
ni	= (k)ni(t), prednisolone
so	= so(n), prednisolone
lone	= lone, (a)lone, (c)lone, triamcinolone

A severe cough or inflammation warrants a liquid steroid like **prednisolone**. Take the p-r-e-d, pred from predator. The n-i, ni is in k-n-i-t, knit. Next, s-o, so isn't like "so what," but has the schwa sound as in s-o-n, son. Finally, l-o-n-e, lone means a single person. Predator, nit, son, lone. Lone. So lone. Ni so lone. Pred ni so lone. Prednisolone.

TEAGAN – A DOUBLE EAR INFECTION

The penicillin antibiotic we are most familiar with is amoxicillin, the pink stuff the pharmacist reconstitutes from powder by adding water. Amoxicillin is a staple for middle ear infections, though some prescribers prefer to let them resolve without treatment. Teagan had an adenoidectomy and tonsillectomy, which means that they took her adenoids and tonsils out. The surgeon put her on amoxicillin immediately post-surgery, yet a week later she started picking at her ears. So, I took her to her pediatrician and she had an ear infection in both ears. Well, if she's already on amoxicillin, what do we do?

My wife and I are pharmacists. We know that if amoxicillin fails, a new therapy would include amoxicillin with clavulanate, brand Augmentin, or a 3rd generation cephalosporin like cefdinir. Both treatments kill enzyme resistant bacteria. Learning antibiotic names helps you have better conversations with your pediatrician, as a pharmacist would.

Penicillin antibiotic

5. Amoxicillin (Amoxil)

a	= (comm)a, (aur)a, acetaminophen
mox	= mox(ie), amoxicillin
i	= i(t), simethicone
cil'	= (pen)cil, amoxicillin
lin	= lin(t), amoxicillin

With **amoxicillin**, brand **Amoxil**, we begin with the "a" from comma — the schwa sound, u-h, uh. M-o-x, mox comes from moxie. Teagan has moxie — she's determined. As I mentioned earlier, the "x" on the interior of a drug name sounds like "k-s" and I was going to keep the consonant "x" on all sounds. Ending "mox" with a consonant leaves the "i" alone and, in a dictionary, it's a schwa sound. I'm recommending we use the "i" from "i-t, it" to make sure we spell amoxicillin right. The c-i-l, cil comes from pencil not pen, and l-i-n, lin from the lint in the clothes dryer. Comma, moxie, it, pencil, lint. Lin. Cil lin. I cil lin. Mox i cil lin. A mox i cil lin. Amoxicillin.

Penicillin antibiotic/Beta-lactamase inhibitor

6-7. Amoxicillin/Clavulanate (Augmentin)

a	= (comm)a, (aur)a, acetaminophen
mox	= mox(ie), amoxicillin
i	= i(t), simethicone

cil'	= (pen)cil, amoxicillin
lin	= lin(t), amoxicillin
cla'	= cla(w), clavulanate
vu	= (re)vu(e), clavulanate
la	= la(wn), la(w), clavulanate
nate	= (in)nate, Nate, calcium carbonate

Clavulanate protects agains bacterial beta-lactamase, an enzyme that renders amoxicillin ineffective. The c-l-a, cla comes from an eagle's claw; the v-u, vu comes from theater revue; the l-a, la is from the lawn in your front yard; and nate comes as part of innate. Claw, revue, lawn, innate. Nate. La nate. Vu la nate. Cla vu la nate. Clavulanate. **Amoxicillin** and **clavulanate** combine to form the brand **Augmentin**.

Cephalosporin antibiotics [by generation]

8. Cephalexin (Keflex) [1st generation]

ce	= ce(nt), ce(nter), cetirizine
pha	= (al)pha, cephalexin
lex'	= (f)lex, cephalexin
in	= in, (t)in, polymyxin B

9. Cefdinir (Omnicef) [3rd generation]

cef'	= c(l)ef, c(l)ef(t), cefdinir
di	= di(nner), di(n), cefdinir
nir	= (souve)nir, cefdinir

Cephalosporins are an antibiotic class whose generic names begin with c-e-p-h or c-e-f. With Teagan, the pediatrician prescribed cefdinir, which would resist the bacteria's beta-lactamase enzyme and quickly resolved her double ear infection.

Cephalexin, brand **Keflex**, begins with c-e, ce from cent, and then comes p-h-a, pha from the Alpha, the first letter in the Greek alphabet. L-e-x, lex comes from f-l-e-x, as I flex muscles, and i-n, in is the opposite of out. Cent, alpha, flex, in. In. lex in. Pha lex in. Ce pha lex in. Cephalexin.

In **cefdinir**, brand **Omnicef**, the c-e-p-h spelling changes to c-e-f, to avoid confusion. The c-e-f, cef comes from the treble clef without the "l" on a music sheet. The d-i, di comes from dinner, a meal after lunch. The n-i-r, nir comes from the end of souvenir. Clef, dinner, souvenir. Nir. Di nir. Cef di nir. Cefdinir.

Macrolide antibiotic

10. Azithromycin (Zithromax)

a	= (comm)a, (aur)a, acetaminophen
zi	= zi(pper), zi(t), azithromycin
thro	= thro(w), azithromycin
my'	= my, my(osin), neomycin sulfate
cin	= cin(der), (s)cin(tillate), bacitracin

Azithromycin, brand **Zithromax** is a macrolide antibiotic with a double dose on the first day, a loading dose. The "a" is the schwa from comma and the z-i, zi comes from clothing zipper. T-h-r-o, thro as in to throw a ball without the "w;" the m-y, my is itself a word; and c-i-n, cin is a fireplace cinder. Comma, zipper, throw, my, cinder. Cin. My cin. Thro my cin. Zi thro my cin. A zi thro my cin. Azithromycin completes a short list of Teagan's meds.

BRIELLE - FENTANYL

Brielle has been brave. In the NICU, she had aortic patent ductus arteriosus, a hole in her blood vessel that didn't close when she left the womb, pyloric valve stenosis where my wife and I had to rescue her with CPR, gluten intolerance, and a need for growth hormone.

Opioid analgesics – Schedule II

11. Fentanyl (Duragesic, Sublimaze)

fen'	= fen, fen(der), ibuprofen
ta	= (da)ta, (sona)ta, (quo)ta, acetaminophen
nyl	= (vi)nyl, fentanyl

After Brielle's surgery, she needed **fentanyl**, an opioid analgesic that we dose in micrograms rather than milligrams. That they gave her, a nine-pound three-month-old a Schedule II opioid, surprised me. To pronounce fentanyl, we use f-e-n, fen, the swamp; the t-a, ta from data; and n-y-l, nyl from vinyl record. Nyl. Ta nyl. Fen ta nyl. Fentanyl.

BRIELLE – ONDANSETRON AND SEVOFLURANE

Antiemetic – Serotonin 5-HT3 receptor antagonist

12. Ondansetron (Zofran)

on	= on, (w)on(ton), ondansetron
dan'	= dan(ce), Dan, ondansetron
se	= se(t), se(nt), ondansetron
tron	= (elec)tron, ondansetron

General inhalation anesthetic – Halogenated alkane derivative

13. Sevoflurane (Sojourn)

se	= se(t), se(nt), ondansetron
vo	= vo(te), sevoflurane
flur'	= flur(ry), sevoflurane
ane	= (c)ane, sevoflurane

We are going to fast-forward as Brielle struggles to grow. She is five years old and 29 pounds. We thought gluten might be the cause, but to get a definitive gluten allergy diagnosis, a gastroenterologist had to biopsy her.

Brielle needed **ondansetron** for her post-op nausea and sevoflurane to put her under. Ondansetron is a strong anti-nausea medication that begins with o-n, on as in turn on the light, and then d-a-n, dan comes from dance. The s-e, se comes from tennis set, and t-r-o-n, tron from electron, the proton's negatively charged partner. On, dance, set, electron. Tron. Se tron. Dan se tron. On dan se tron. Ondansetron.

Chapter B - Prescription children's medications

Sevoflurane, brand **Sojourn**, the inhaled anesthetic, uses the s-e, se from set, the same as ondan-se-tron, v-o, vo from vote, f-l-u-r, from flurry of snowflakes, and a-n-e, ane from wooden cane. Set, vote, flurry, cane. Ane. Flur ane. Vo flur ane. Se vo flur ane. Sevoflurane.

Brielle had an elevated liver function test common to celiac disease and the gastroenterologist wanted to make sure, so he needed to perform a biopsy.

The night before the procedure, 30 minutes until midnight, I get Brielle up to eat. She's having an endoscopy done in 11 hours and she can't eat for the 8 hours before. I've got to get her Pediasure. We'll go downstairs to the refrigerator.

"Daddy, I want to pour it in myself," she says.

"First, chocolate. Then vanilla," she reminds me as I hand her the cans.

Afterward, I lay her back down, and she says, "Daddy, sleep by me."

At the hospital, my wife texts me that the planet Mercury is visible in front of the sun at the Iowa Science Center from 10 AM to 1 PM. This is her way of asking for text updates, as it's overcast with no sun outside.

I'm listening to *Lab Girl* on my iPhone, an autobiography of a Ph.D. scientist.

We get directions to the waiting room: right at the chapel, left at the pop machine. Two women in the waiting room pray. The receptionist isn't here yet. One woman says, "C'mon girl, where are you?"

Brielle says, "My shoes aren't staying on, they're too big." I've realized we've been here before for her GERD esophagitis. "Here I Go Again" by Whitesnake plays on the radio. Seriously?

Distracted, on the intake form under Brielle's psychosocial needs, I write "her puppy and iPad."

The impatient woman comes back and asks the volunteer receptionist with green scrubs to start some coffee. Her friend reminds her that she can't have any. Now a coffee aroma taunts the fasting patients.

My mind wanders. Green scrubs. Why do people stop at green arrows? Why do I duck when I drive under yellow lights?

I take Brielle in and ask the anesthesiologist about the anesthetic and whisper what he said to Brielle. He'll use sevoflurane, brand name Sojourn. He says to Brielle, "it smells like a dinosaur's armpit." He's kind. A sojourn, the brand name, is a temporary stay. Brand name drugs often speak as one word poems.

We go in and, as I hand her off, she wraps her tiny hand around my finger. The anesthesiologist rocks her in his arms like a toddler on Santa's lap and places a mask over her nose and mouth. She'll fight sleep, then slowly lose her grip on my hand and pass out. I'll slowly lose my grip on holding it together as a nurse walks me back to the waiting room.

Ondansetron prevents her nausea and she recovers quickly. She'll wake up, have an apple juice box, and eat my last peanut butter CLIF bar. I'll dress her while she watches her iPad, and then head to the VA hospital to see Mom at work.

All of this has been overhwhelming, but understanding the drugs — and their effects and uses — means a great deal when it comes to finding the right treatment for your child, and keeping your sanity when stressed about their safety and health. Understanding, in a case like this, actually makes for some small amount of comfort.

Anesthetics

14. Lidocaine (Lidoderm) [Amide type]

li'	= li(ght), li(e), lidocaine
do	= do(ugh), lidocaine
caine	= c(at)(migr)aine, benzocaine

Lidocaine, a local numbing anesthetic, comes in various forms. For Brielle, they used a patch to numb her arms for an IV needle. Lidocaine begins with l-i, li from light bulb or white lie; d-o, do as the first two letters and with the same sound as cookie dough; and with c-a-i-n-e, caine, we repeat what we did with benzocaine. Start with "c" from cat and add a-i-n-e, aine from migraine. Light, dough, cat, migraine. Caine. Do caine. Li do caine. Lidocaine.

15. Propofol (Diprivan)

pro'	= pro(ton), pro(be), pro(active), ibuprofen
po	= po(tato), propofol
fol	= fol(licle), fol(ly), propofol

Intead of sevoflurane for the MRI, the anesthesiologist injected **propofol**, brand **Diprivan**. The p-r-o, pro from proton with the p-o, po from potato with the "o" taking the schwa u-h, uh sound and f-o-l, fol from hair follicle. Proton, potato, follicle. Fol. Po fol. Pro po fol. Propofol.

Peptide hormone

16. Somatropin (Norditropin)

so	= so(fa), so, docusate sodium
ma	= (paja)ma, somatropin
tro'	= (me)tro, (re)tro, dextromethorphan
pin	= pin, somatropin

Somatropin, brand **Norditropin**, starts with the s-o, so from sofa; goes to m-a, ma from pajama, with the "a" taking the schwa sound; t-r-o, tro from metro; and p-i-n, pin like safety pin. Sofa, pajama, metro, pin. Pin. Tro pin. Ma tro pin. So ma tro pin. Somatropin.

BRIELLE – LIDOCAINE, PROPOFOL, SOMATROPIN

Lidocaine

Our endocrine specialist told us that, to have the insurance cover growth hormone, we would need to have Brielle complete an insulin stress test. They would give her insulin through an IV in her arms to push her glucose down, and this would provide the evidence the insurance company needed.

Before they inserted the IV, they put two lidocaine patches on her arms so she wouldn't feel it. They tried the left arm, and then the right with no success. The nurses recommended we try her hand. I said, "What about the lidocaine?" The nurses believed the lidocaine would mask the vein. If you make a fist, you can see what they were shooting for, but without anesthesia, the piercing "owie, owie, owie" when they pushed the needle in her unanaesthetized hand . . . it broke us. We stopped the procedure. We took her to the gift shop to get her a candy and a toy, and then took her home. Her limp arm reminded me I'd failed as a parent.

The procedure was unnecessary. Some insurance agencies required the insulin stress test, but based on her birth weight, and her progress on the growth chart under the first percentile, she was eligible for consideration for the growth hormone without the stress test.

I have years of experience with insurance. I confused clinical expertise with administrative competence. I assumed the nurse called our insurance, but instead she'd relied on her previous experience.

Note: This is why I'm so excited about IBM's Watson entering the health care field. Most people talk about Watson and its ability to work with clinicians, but supercomputer like Watson could also review insurance conditions and prevent this event. I'll never forgive myself for okaying that needle push into her fist.

Propofol
After getting the okay from the insurance, we still needed to complete Brielle's MRI to insure a tumor wasn't causing her growth issue. Understandably, the night before, she said, "Daddy, I want to whisper something in your ear."

Chapter B - Prescription children's medications

"Sure, what is it?" I asked.

"Daddy, tell the doctors no needle."

No such luck, as she was getting propofol.

The nurses, the life flight crew, have Navy blue flight suits and dark blue vests. This time, they've brought the best in pediatric care to help us. I see their ear buds for helicopter noise and keep associating bubbles with needles. It's a white viscous syringe of propofol. Brielle hugs her pink pony.

"Daddy, no needles, no needles."

She screams through tears, "I want to go home right now," and "owie, owie, owie, owie."

The anesthesiologist, she says, "I'm sorry, it burns."

Brielle screams, "I want to be done, I want to go home!" as her pink pony falls and she begins to lose consciousness.

They let us put the guardrail down for a moment, and my wife and I give her a kiss. She has a Pedicraft IV white board attached to her hand and forearm. They take her away and we wait to find out if our daughter has a tumor.

Somatropin
The MRI is clear. We get the green light for growth hormone, the somatropin, and still question our decision. We've read the scientific literature, but it comes down to bits of foreshadowing we both saw and heard.

Brielle said, "Since I'm the littleist, I have to stand in front, some people say."

Now, Brielle gets daily injections after she falls asleep. She squirms from the BD Nano needle injections, and then usually drops back into REM sleep.

You can tell she wants to grow. In the family room, she took all of her shoes out of her bin. Then, one by one, she took a Stanley tape measure to them. She extended the measuring tape parallel to each shoe, to mark her progress growing.

RIANNE – SULFAMETHOXAZOLE AND TRIMETHOPRIM

Dihydrofolate reductase inhibitors

17. Sulfamethoxazole / Trimethoprim (Bactrim DS)

sul	= sul(len), neomycin sulfate
fa	= (so)fa, sulfamethoxazole/trimethoprim
meth	= meth(od), simethicone
ox'	= ox, (b)ox, sulfamethoxazole/trimethoprim
a	= (comm)a, (aur)a, acetaminophen
zole	= (fe)z(p)ole, sulfamethoxazole/trimethoprim
tri	= tri(angle), triamcinolone
meth'	= meth(od), simethicone
o	= "o," fexofenadine
prim	= prim, prim(p), sulfamethoxazole/trimethoprim

Sulfamethoxazole and **trimethoprim** both block folic acid production by killing bacteria, especially those responsible for urinary tract infections. Providers often shorten it to SMZ/TMP, taking six of the two drugs' letters.

To pronounce sulfamethoxazole, we take s-u-l, sul from sullen; f-a, fa from sofa; and m-e-t-h, meth from method. Add o-x, ox, the beast of burden and "a" from comma. To pronounce z-o-l-e, zole, combine the "z" from fez, the red Turkish hat the organ grinder monkey wears, and o-l-e, ole from pole. Sullen, sofa, method, ox, comma, fez, pole. Zole. A zole. Ox a zole. Meth ox a zole. Fa meth ox a zole. Sul fa meth ox a zole. Sulfamethoxazole.

Trimethoprim takes the t-r-i, tri from triangle; the m-e-t-h, meth, from method; "o" as in the letter "o;" and p-r-i-m, prim from the phrase "prim and proper." Triangle, method, o, prim. Prim. O prim. Meth o prim. Tri meth o prim. Trimethoprim.

As I sit in the dark of exam room three, I ask my one-year-old daughter Rianne, "Am I a good father?" Her fever burns my arm, tears wet my shirt, labored breathing ripples the fabric between my ribs, and her eyelids open and then close, blinking out blue streams of luciferin light. Rianne came in with what we thought was a urinary tract infection.

I held her down but, as soon as the doctor started trying to catheterize her, Rianne urinated all over us. I think Rianne did it on purpose, to be honest; it's like a bird that flies over your head and drops a bomb. Are those birds aiming? Rianne was definitely ill, but if I didn't know better, she was smiling. She got a prescription for sulfamethoxazole and trimethoprim, an antibiotic to clear the urinary tract infection.

MISCELLANEOUS MEDICATIONS

These other medications fit less neatly into a story. First, though, I want to tell you about a memorable NICU moment – the triplet kangaroo. Kangarooing is having your NICU child lay on your chest to take advantage of your warmth, heart, and respiratory rate.

TRIPLET KANGAROO

My triplet daughters, naked but for diapers, meet for the first time on my bare chest as I kangaroo them in this sticky vinyl recliner. Two nurses hover nearby, admiring their handiwork as oxygen tubes and colored leads measuring heart and respiratory rates run from the incubators. My wife snaps digital pictures. Brielle suckles for a nipple from which to extract milk. Good luck. Teagan pushes, then punches Rianne away while fighting for a more comfortable bunk. A nurse dons each girl with a knit hat and two blankets, and exits. The lights dim. I close my eyes and feel one of my daughters nuzzle her head, like a cat circling before a nap. Another rips out a chest hair. Their vitals improve as they calm themselves by aligning with my heart and breathing rate. I feel like Alan Shepard in a cramped space capsule in *The Right Stuff*. I let myself fall away, sweating from the heat, sweating as a new father. I whisper Alan Shepard's prayer, "Dear Lord, please don't let me f--- up."

Macrolide antibiotic

18. Ery<u>thromycin</u> (E-Mycin)

e	= (en)e(my), erythromycin
ry	= (o)ry(x), (t)ry(st), erythromycin
thro	= thro(w), azithromycin
my'	= my, my(osin), neomycin sulfate
cin	= cin(der), (s)cin(tillate), bacitracin

Like azithromycin, brand Zithromax, **erythromycin**, brand **E-Mycin**, is a macrolide antibiotic. The first "e" takes the schwa, as in the middle "e" in enemy. The r-y, ry comes from oryx, an antelope. The last three syllables are identical to the azithromycin ending with t-h-r-o, m-y, and c-i-n, from throw, my, and cinder respectively. Enemy, oryx, throw, my, cinder. Cin. My cin. Thro my cin. Ry thro my cin. E ry thro my cin. Erythromycin.

Topical antibiotic

19. Mupi<u>ro</u>cin (Bactroban)

mu	= mu(sic), mu(seum), mupirocin
pi'	= pi(ece), ipratropium
ro	= ro(w), ro(ad), mupirocin
cin	= cin(der), (s)cin(tillate), bacitracin

Mupirocin, brand **Bactroban**, is for skin infections like impetigo. The m-u, mu comes from the Greek letter or the word music; the p-i, pi sound stems from a puzzle piece; the r-o, ro is like row a boat; and c-i-n, cin comes from cinder. Music, piece, row, cinder. Cin. Ro cin. Pi ro cin. Mu pi ro cin. Mupirocin.

Non-narcotic analgesic with opioid analgesic - Schedule V

20-21. Acetaminophen w/codeine (Tylenol/codeine)

a	= (comm)a, (aur)a, acetaminophen
ce	= ce(iling), acetaminophen
ta	= (da)ta, (sona)ta, (quo)ta, acetaminophen
mi'	= mi(nt), mi(lk), acetaminophen
no	= (can)no(n), (ca)no(n), acetaminophen
phen	= (hy)phen, acetaminophen
co'	= co(ne), codeine
deine	= de(er), (mach)ine, codeine

We've gone over **acetaminophen**, an analgesic, but **codeine** is a good antitussive, a drug for cough. Take the c-o, co from an ice cream cone to start, and then combine two words to get the d-e-i-n-e, deine sound. It's snowing in Ankeny, Iowa, so I thought of the d-e, de from d-e-e-r, deer and i-n-e, ine from machine. Cone, deer, machine. Deine. Co deine. Codeine.

NICU MEDS

22. Surfactant

sur	= sur(fer), sur(f), surfactant
fac'	= fac(t), fac(tor), surfactant
tant	= (sex)tant, surfactant

Surfactant is a medicine they gave our daughters at birth to help their lungs. Our triplets were born at 27 weeks, at 1 ½, 1 ½, and 2 pounds. The image I have is of the neonatologist rolling baby one to the left, then right, rolling baby two to the left and then right, and then there was baby three.

Take the s-u-r, sur from surfer dude; f-a-c, fac from fact, as in fact not fiction; and t-a-n-t, tant from sextant, a nautical navigation instrument. Surfer, fact, sextant. Tant. Fac tant. Sur fac tant. Surfactant.

23. Caffeine

| caf' | = caf(e), caffeine |
| feine | = fe(et)(magaz)ine, caffeine |

Caffeine helps with apnea, or the tendency for the babies to stop breathing in the NICU. Take the c-a-f, caf from café, and combine the f-e, fe from feet with the i-n-e, ine from magazine. Enjoy coffee with caffeine at the café while you put your feet up and read a magazine. Café, feet, magazine. Feine. Caf feine. Caffeine.

Glycopeptide antibiotic

24. Vancomycin (Vancocin)

van	= van, van(dal), vancomycin
co	= co(ne), codeine
my'	= my, my(osin), neomycin sulfate
cin	= cin(der), (s)cin(tillate), bacitracin

Aminoglycoside antibiotic

25. Gentamicin (Garamycin)

gen	= gen(tleman), gentamicin
ta	= (da)ta, (sona)ta, (quo)ta, acetaminophen
mi'	= mi(le), gentamicin
cin	= cin(der), (s)cin(tillate), bacitracin

We pronounce the last two syllables to **vancomycin**, v-a-n-c-o-m-y-c-i-n, and **gentamicin**, g-e-n-t-a-m-i-c-i-n, the same. However, the spellings m-y-c-i-n and m-i-c-i-n indicate that the antibiotics originate from different bacteria. Often, bacteria produce the antibiotics we use to kill other bacteria.

Vancomycin includes v-a-n, van as in minivan; c-o, co from ice cream cone; m-y, my as the word itself; and c-i-n, cin from cinder. Van, cone, my, cin. Cin. My cin. Co my cin. Van co my cin. Vancomycin.

Gentamicin takes the g-e-n, gen from gentleman; the t-a, ta from data; the m-i, mi from mile; and the c-i-n, cin from cinder. Gentleman, data, mile, cinder. Cin. Mi cin. Ta mi cin. Gen ta mi cin. Gentamicin.

Respiratory syncytial virus (RSV)

26. Palivizumab (Synagis)

pa	= pa(th), palivizumab
li	= li(ly), li(nt), palivizumab
vi'	= vi(sion), vi(m), palivizumab
zu	= Zu(lu), (shiat)zu, palivizumab
mab	= m(at)(c)ab, palivizumab

RSV, or respiratory syncytial virus, is a dangerous infection. As adults with competent immune systems, we don't worry much about it, but those weak neonates must. **Palivizumab**, brand **Synagis**, is the preventative medication. We stored it in the refrigerator. However, it's so expensive that a nurse drives to our house, measures it, calls in her dosages, and then injects it.

Palivizumab starts with p-a, pa from path; moves to the l-i, li from lily; the v-i, vi from vision; the z-u, zu from Zulu, a South African tribe; and then the "m" from m-a-t, mat and the a-b, ab from taxi cab. Path, lily, vision, Zulu, mat, cab. Mab. Zu mab. Li zu mab. Pa li vi zu mab. Palivizumab.

Influenza virus vaccine

27. Influenza vaccine (Fluzone, various others)

in	= in, (t)in, polymyxin B
flu	= flu(id), fluticasone
en'	= (p)en, (m)en, influenza vaccine
za	= (pla)za, influenza vaccine
vac'	= vac(uum), influenza vaccine
cine	= (leu)cine, influenza vaccine

Influenza vaccine is a source of great dismay for our girls each Fall. I took them all myself this year and the girls were surprisingly aware. They elected a first, second, and third volunteer. Rianne, Teagan, Brielle. I was really proud of them, and what they said: "If you don't get your flushot, then you're gonna get sick."

Influenza breaks down to i-n, in from in and out; f-l-u, flu from fluid; e-n, en from pen; and z-a, za from plaza with the schwa "a" sound. In, fluid, pen, plaza. Za. En za. Flu en za. In flu en za. Influenza.

Vaccine comes from v-a-c-a, vaca which is Latin for cow. I used the v-a-c, vac from vacuum cleaner and the c-i-n-e, cine from leucine, an amino acid. Vacuum, leucine. Cine. Vac cine. Vaccine.

Chapter C - Recording
Pronunciations in Your Own Voice

In this chapter, I am going to talk to you about recording the pronunciations in your own voice so that you can match them against my YouTube videos on the TonyPharmD YouTube channel. That's T-o-n-y-P-h-a-r-m-D, my name and my professional degree as one word. You've learned over 50 medication names so far, but without testing yourself, how do you know you are pronouncing them correctly?

I believe drug names come alive when you can hear yourself saying them and know they are correct. This is also a chance for teacher and student, parent and child, or parent and parent to work together. Patients can be more confident in their contributions to the electronic health record (EHR), as well, such that there's meaningful use. Meanginful use is working to improve quality and reduce health inequities for patients and families, especially as they relate to public health.

Let's work to pronounce the common urinary tract infection medication, sulfamethoxazole and trimethoprim. It's 10 syllables and intimidating. Here is the pronunciation scheme.

sul	= sul(len), neomycin sulfate
fa	= (so)fa, sulfamethoxazole/trimethoprim
meth	= meth(od), simethicone
ox'	= ox, (b)ox, sulfamethoxazole/trimethoprim
a	= (comm)a, (aur)a, acetaminophen
zole	= (fe)z(p)ole, sulfamethoxazole/trimethoprim
tri	= tri(angle), triamcinolone
meth'	= meth(od), simethicone
o	= "o", fexofenadine
prim	= prim, prim(p), sulfamethoxazole/trimethoprim

In a linguistics class, so that we don't let an adjacent word affect the sound, we'll pronounce each word into the microphone by including a small phrase like, "Now I say." Pause between "say", the target syllable, and "from".

Script

Now I say ... sul ... **from "sullen"**

Now I say ... fa ... **from "sofa"**

Now I say ... meth ... **from "method"**

Now I say ... ox ... **from "ox"**

Now I say ... a ... **from "comma"**

Now I say ... zole ... **from "fez" and "pole"**

Now I say ... tri ... **from "triangle"**

Now I say ... meth ... **from "method"**

Now I say ... "o" ... **from "o"**

Now I say ... prim ... **from "prim"**

Now I say

Look at the voiceprint or spectrogram. You'll see that each "Now I say" imprint and "from" imprint look similar visually.

1. Delete the first "Now I say"
2. Delete all instances of "from" to the end of "say"
3. Listen to what's left. It's the drug name in your own voice

I used a program called Sound Forge. I've also been able to use P-R-A-A-T, PRAAT, a free computer software program for acoustic analysis.

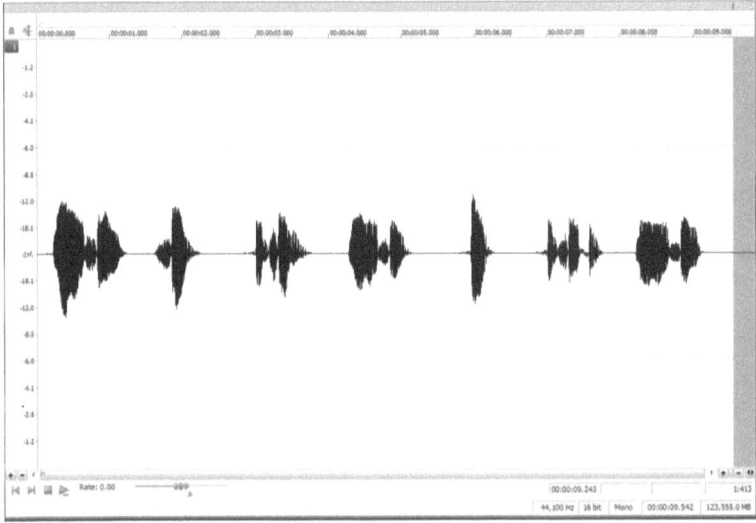

Figure 1- Recording without any edits

This is a part of the recording of the entire script where we can see the sound file visually. I'm going to first label the sounds so that you can see what each of them represents in Figure 2.

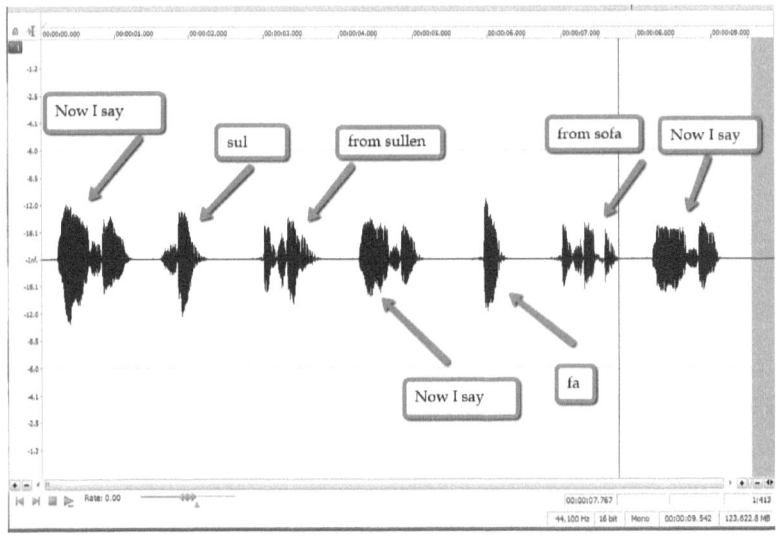

Figure 2 - I've labeled each sound so you can see what I'm doing

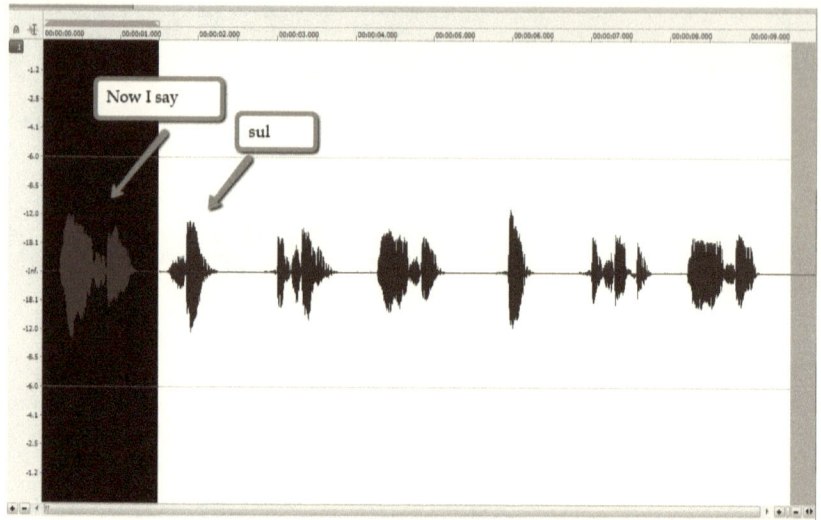

Figure 3 - Preparing to cut

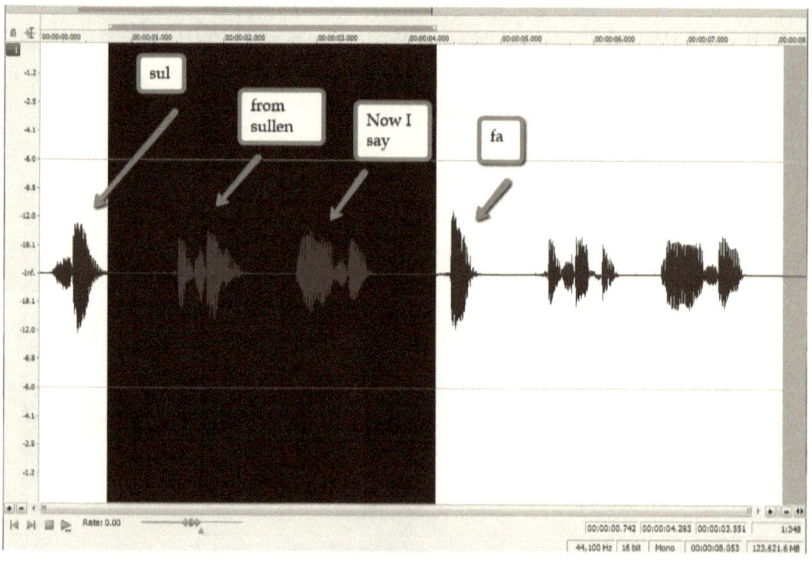

Figure 4 - I've cut the first "Now I say" and now I'm going to cut "from" to "say"

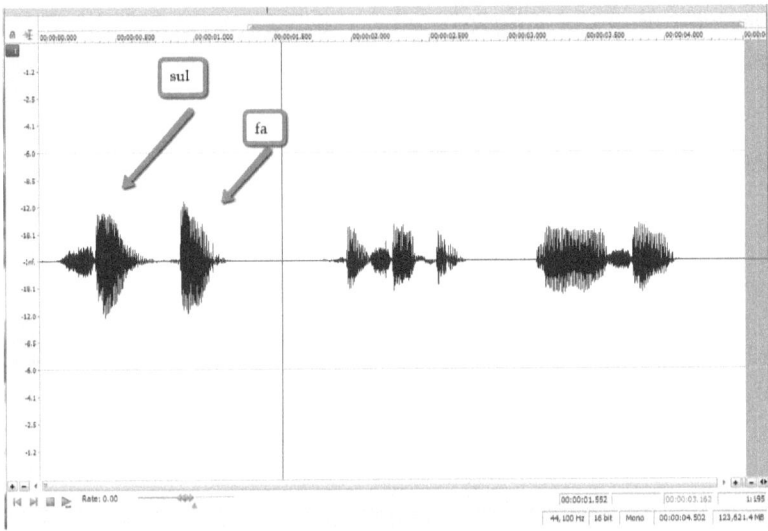

Figure 5 - Now I'm left with the first two syllables of the drug name

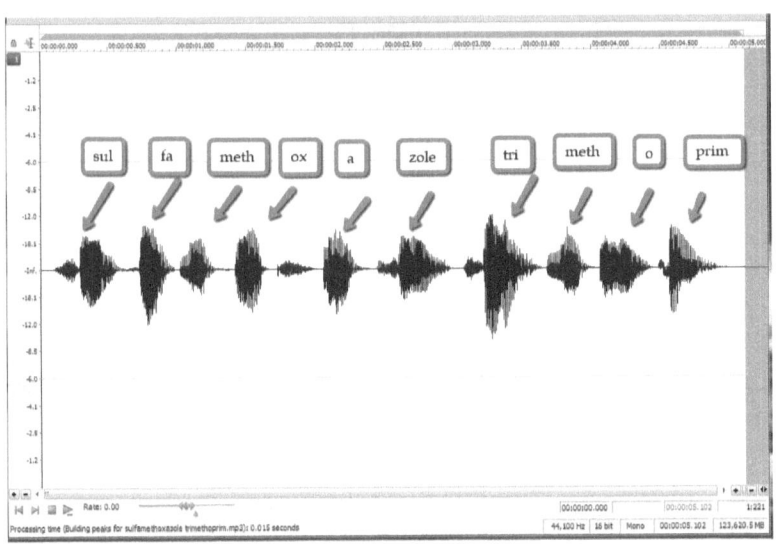

Figure 6 - After you remove all of the "from" to "say" sections, you are left with the drug name in your own voice as an .mp3

I believe there could be some utility in being able to use English words from my system in the Electronic Health Record (EHR) because you then don't have to worry about the computer's accent. Instead, a person has it in his or her own English dialect.

HUMAN-COMPUTER INTERACTION

Should people find the Generic to English Translation System (GETS) an effective tool that improves usability, I can create common English word translations for all of the 20,000 drugs in the United States Pharmacopeia (USP). Or, I can collaborate with someone at the British National Formulary, the BNF, or the NPS MedicineWise group in Australia or MIMS, the monthly index of medical specialties. GETS is flexible with different English dialects. The next step is to work with others that enjoy improving human-computer interaction, or HCI.

CHAPTERS 1 THROUGH 7 [350 DRUGS]

The next seven chapters contain 350 top drugs from print and eBook editions of *Memorizing Pharmacology: A Relaxed Approach*. I've gone through each medication and broken it into common English words, which I've put into the Generic to English Translation System list. After these 7 chapters, you can use the list to predict what new medications, such as the pediatric oncology medications in Chapter 8, will sound like. I've underlined stems to help you recognize the cues that a medication belongs to a particular drug class. If readers want expanded descriptions for all 350, then I'll also be working to do that in the future. I didn't put the backbuilding here, but remember the tool is there if you need it.

CHAPTER 1 - GASTROINTESTINAL

MEDICATIONS

I. PEPTIC ULCER DISEASE

Antacids

1. Calcium Carbonate (Tums)

cal'	= cal(orie), (lo)cal(e), calcium carbonate
ci	= (pronun)ci(ation), calcium carbonate
um	= (g)um, calcium carbonate
car'	= car, car(d), calcium carbonate
bo	= bo(ne), calcium carbonate
nate	= (in)nate, Nate, calcium carbonate

2. Magnesium Hydroxide (Milk of Magnesia)

mag	= mag(azine), magnesium hydroxide
ne'	= (mo)ne(y), k(ne)e, magnesium hydroxide
si	= (enthu)si(astic), (tran)si(ent), magnesium hydroxide
um	= (g)um, calcium carbonate
hy	= hy(brid), hy(phen), magnesium hydroxide
drox'	= dr(ink)(b)ox, dr(op)(b)ox, magnesium hydroxide
ide	= (sl)ide, magnesium hydroxide

Histamine-2 Receptor Antagonists (H₂RAs)

3. Famotidine (Pepcid)

fa	= (so)fa, sulfamethoxazole/trimethoprim
mo'	= mo(at), famotidine
ti	= ti(c), (plas)ti(c), cetirizine
dine	= (bir)di(e)(tu)ne, fexofenadine

4. Rani<u>tidine</u> (Zantac)

ra	= (au)ra, (cob)ra, (Op)ra(h), loperamide
ni'	= (k)ni(t), prednisolone
ti	= ti(c), (plas)ti(c), cetirizine
dine	= (bir)di(e)(tu)ne, fexofenadine

Proton Pump Inhibitors (PPIs)

5. Dexlanso<u>prazole</u> (Dexilant)

dex	= dex(terity), Dex(ter), dextromethorphan
lan	= lan(ce), lan(d), dexlansoprazole
so'	= so(fa), so, docusate sodium
pra	= (su)pra, (O)pra(h), dexlansoprazole
zole	= (fe)z(p)ole, sulfamethoxazole/trimethoprim

6. Esome<u>prazole</u> (Nexium)

es'	= (m)es(s), Es(ther), esomeprazole
o	= "o", fexofenadine
me'	= me(n), me(sh), esomeprazole
pra	= (su)pra, (O)pra(h), dexlansoprazole
zole	= (fe)z(p)ole, sulfamethoxazole/trimethoprim

7. Ome<u>prazole</u> (Prilosec)

o	= "o", fexofenadine
me'	= me(n), me(sh), esomeprazole
pra	= (su)pra, (O)pra(h), dexlansoprazole
zole	= (fe)z(p)ole, sulfamethoxazole/trimethoprim

8. Lanso<u>prazole</u> (Prevacid)

lan	= lan(ce), lan(d), dexlansoprazole
so'	= so(fa), so, docusate sodium
pra	= (su)pra, (O)pra(h), dexlansoprazole
zole	= (fe)z(p)ole, sulfamethoxazole/trimethoprim

9. Panto<u>prazole</u> (Protonix)

pan	= pan, pantoprazole
to'	= to(e), to(w), pantoprazole
pra	= (su)pra, (O)pra(h), dexlansoprazole

zole   = (fe)z(p)ole, sulfamethoxazole/trimethoprim

10. Rabeprazole (AcipHex)

ra   = (au)ra, (cob)ra, (Op)ra(h), loperamide
be'   = be(d), rabeprazole
pra   = (su)pra, (O)pra(h), dexlansoprazole
zole   = (fe)z(p)ole, sulfamethoxazole/trimethoprim

II. DIARRHEA, CONSTIPATION, AND EMESIS

Antidiarrheals

11. Bismuth Subsalicylate (Pepto-Bismol)

bis'   = bis(quit), Bis(marck), bismuth subsalicylate
muth   = (azi)muth, bismuth subsalicylate

sub   = sub(marine), sub(way), bismuth subsalicylate
sa   = (vi)sa, sa(liva), bismuth subsalicylate
li'   = li(ly), li(nt), palivizumab
cy   = cy(st), bismuth subsalicylate
late   = (p)late, bismuth subsalicylate

12. Loperamide (Imodium)

lo   = (Co)lo(rado), (s)lo(w), loperamide
pe'   = pe(n), pe(ril), pe(t), loperamide
ra   = (au)ra, (cob)ra, (Op)ra(h), loperamide
mide   = (gu)m(sl)ide, loperamide

13. Diphenoxylate / atropine (Lomotil)

di   = di(ce), diphenhydramine
phe   = phe(nomenon), brompheniramine
nox'   = nox(ious), (equi)nox, diphenoxylate / atropine
yl   = (s)yl(lable), polyethylene glycol
ate   = (g)ate, (l)ate, diphenoxylate / atropine

a'   = (m)a(t), (b)a(t), diphenoxylate / atropine
tro   = (me)tro, (re)tro, dextromethorphan

pine   = (shi)p(mar)ine, atropine

Constipation – Stool softener

14. Docusate sodium (Colace)

do'   = do(t), do(dge), docusate sodium
cu   = cu(be), cu(re), cu(te), docusate sodium
sate   = (pul)sate, sate, (compen)sate, docusate sodium

so'   = so(fa), so, docusate sodium
di   = (bir)di(e), (per)di(em), docusate sodium
um   = (g)um, calcium carbonate

Constipation – Osmotic

15. Polyethylene glycol (PEG) 3350 (MiraLax)

po   = po(t), po(lygon), polyethylene glycol
ly   = (li)ly, (po)ly(gon), polyethylene glycol
eth'   = (m)eth(od), polyethylene glycol
yl   = (s)yl(lable), polyethylene glycol
ene   = (sc)ene, polyethylene glycol

gly'   = gly(cemic), polyethylene glycol
col   = col(lar), Col(orado), polyethylene

Constipation – Miscellaneous

16. Lubiprostone (Amitiza)

lu   = lu(cid), montelukast
bi   = bi(t), (or)bi(t), lubiprostone
pro'   = pro(m), pro(d), lubiprostone
stone   = stone, lubiprostone

Antiemetic – Serotonin 5-HT$_3$ receptor antagonist

17. Ondansetron (Zofran)

on   = on, (w)on(ton), ondansetron
dan'   = dan(ce), Dan, ondansetron

se   = se(t), se(nt), ondansetron
tron   = (elec)tron, ondansetron

Antiemetic – Phenothiazine

18. Prochlorperazine (Compazine)

pro   = pro(ton), pro(be), pro(active), ibuprofen
chlor   = chlor(ine), chlorpheniramine
pe'   = pe(n), pe(ril), pe(t), loperamide
ra   = (au)ra, (cob)ra, (Op)ra(h), loperamide
zine   = (maga)zine, cetirizine

19. Promethazine (Phenergan)

pro   = pro(ton), pro(be), pro(active), ibuprofen
meth'   = meth(od), simethicone
a   = (comm)a, (aur)a, acetaminophen
zine   = (maga)zine, cetirizine

III. AUTOIMMUNE DISORDERS

Ulcerative colitis

20. Infliximab (Remicade)

in   = in, (t)in, polymyxin B
flix'   = fl(our)(m)ix, infliximab
i   = i(t), simethicone
mab   = m(at)(c)ab, palivizumab

CHAPTER 2 - MUSCULOSKELETAL

MEDICATIONS

I. NSAIDs AND PAIN

OTC Analgesics – NSAIDs

21. Aspirin [ASA] (Ecotrin)

as'	= as(p), aspirin
pi	= pi(n), aspirin
rin	= rin(se), permethrin

22. Ibuprofen (Advil, Motrin)

i	= "i", ibuprofen
bu	= bu(reau), ibuprofen
pro'	= pro(ton), pro(be), pro(active), ibuprofen
fen	= fen, fen(der), ibuprofen

23. Naproxen (Aleve)

na	= (bana)na, fexofenadine
prox'	= pr(oton)(b)ox, naproxen
en	= (p)en, (m)en, influenza vaccine

OTC Analgesic – Non-narcotic

24. Acetaminophen [APAP] (Tylenol)

a	= (comm)a, (aur)a, acetaminophen
ce	= ce(iling), acetaminophen
ta	= (da)ta, (sona)ta, (quo)ta, acetaminophen
mi'	= mi(nt), mi(lk), acetaminophen
no	= (can)no(n), (ca)no(n), acetaminophen
phen	= (hy)phen, acetaminophen

OTC Migraine – NSAID / Non-narcotic analgesic

25. ASA/APAP/Caffeine (Excedrin Migraine)

as'	= as(p), aspirin
pi	= pi(n), aspirin
rin	= rin(se), permethrin
a	= (comm)a, (aur)a, acetaminophen
ce	= ce(iling), acetaminophen
ta	= (da)ta, (sona)ta, (quo)ta, acetaminophen
mi'	= mi(nt), mi(lk), acetaminophen
no	= (can)no(n), (ca)no(n), acetaminophen
phen	= (hy)phen, acetaminophen
caf'	= caf(e), caffeine
feine	= fe(et)(magaz)ine, caffeine

RX Migraine – Narcotic and Non-narcotic analgesic

26. Butalbital / APAP / Caffeine (Fioricet)

bu	= bu(reau), ibuprofen
tal'	= tal(l), butalbital
bi	= bi(t), (or)bi(t), lubiprostone
tal	= tal(l), butalbital
a	= (comm)a, (aur)a, acetaminophen
ce	= ce(iling), acetaminophen
ta	= (da)ta, (sona)ta, (quo)ta, acetaminophen
mi'	= mi(nt), mi(lk), acetaminophen
no	= (can)no(n), (ca)no(n), acetaminophen
phen	= (hy)phen, acetaminophen
caf'	= caf(e), caffeine
feine	= fe(et)(magaz)ine, caffeine

RX Analgesics – NSAIDs

27. Diclofenac sodium extended release (Voltaren XR)

di	= di(ce), diphenhydramine
clo'	= clo(ver), clo(the), diclofenac
fe	= fe(n), fexofenadine
nac	= (k)nac(k), diclofenac

28. Etodolac (Lodine)

e	= (n)e(t), etodolac
to'	= to(e), to(w), pantoprazole
do	= (i)do(l), etodolac
lac	= (li)lac, lac(tose), spironolactone

29. Indomethacin (Indocin)

in	= in, (t)in, polymyxin B
do	= do(ugh), lidocaine
meth'	= meth(od), simethicone
a	= (comm)a, (aur)a, acetaminophen
cin	= cin(der), (s)cin(tillate), bacitracin

30. Meloxicam (Mobic)

me	= me(n), me(sh), esomeprazole
lox'	= lox, meloxicam
i	= i(t), simethicone
cam	= cam(p), meloxicam

31. Nabumetone (Relafen)

na	= (bana)na, fexofenadine
bu'	= bu(reau), ibuprofen
me	= me(n), me(sh), esomeprazole
tone	= tone, (s)tone, nabumetone

RX Analgesics – NSAIDs – COX-2 inhibitor

32. Celecoxib (Celebrex)

ce	= ce(nt), ce(nter), cetirizine
le	= le(tter), celecoxib

cox¹   = cox(swain), celecoxib
ib   = (r)ib, celecoxib

II. OPIOIDS AND NARCOTICS

Opioid analgesics – Schedule II

33a. Hydromorphone (Dilaudid)

hy   = hy(brid), hy(phen), magnesium hydroxide
dro   = dro(ne), hydrocortisone
mor¹   = mor(e), Mor(se)(code), hydromorphone
phone   = phone, hydromorphone

33b. Morphine (Kadian, MS Contin)

mor¹   = mor(e), Mor(se)(code), hydromorphone
phine   = (So)phi(a)(du)ne, morphine

34. Fentanyl (Duragesic, Sublimaze)

fen¹   = fen, fen(der), ibuprofen
ta   = (da)ta, (sona)ta, (quo)ta, acetaminophen
nyl   = (vi)nyl, fentanyl

35. Hydrocodone / Acetaminophen (Vicodin)

hy   = hy(brid), hy(phen), magnesium hydroxide
dro   = dro(ne), hydrocortisone
co¹   = co(ne), codeine
done   = (con)done, hydrocodone

a   = (comm)a, (aur)a, acetaminophen
ce   = ce(iling), acetaminophen
ta   = (da)ta, (sona)ta, (quo)ta, acetaminophen
mi¹   = mi(nt), mi(lk), acetaminophen
no   = (can)no(n), (ca)no(n), acetaminophen
phen   = (hy)phen, acetaminophen

36. Hydrocodone / Chlorpheniramine (Tussionex)

hy   = hy(brid), hy(phen), magnesium hydroxide

dro	= dro(ne), hydrocortisone
co'	= co(ne), codeine
done	= (con)done, hydrocodone

chlor	= chlor(ine), chlorpheniramine
phe	= phe(nomenon), brompheniramine
ni'	= ni(ght), brompheniramine
ra	= (au)ra, (cob)ra, (Op)ra(h), loperamide
mine	= (hista)mine, brompheniramine

37. Hydrocodone / Ibuprofen (Vicoprofen)

hy	= hy(brid), hy(phen), magnesium hydroxide
dro	= dro(ne), hydrocortisone
co'	= co(ne), codeine
done	= (con)done, hydrocodone

i	= "i", ibuprofen
bu	= bu(reau), ibuprofen
pro'	= pro(ton), pro(be), pro(active), ibuprofen
fen	= fen, fen(der), ibuprofen

38. Methadone (Dolophine)

meth'	= meth(od), simethicone
a	= (comm)a, (aur)a, acetaminophen
done	= (con)done, hydrocodone

39. Oxycodone (OxyIR, Oxycontin)

ox	= ox, (b)ox, sulfamethoxazole/trimethoprim
y	= (countr)y, oxycodone
co'	= co(ne), codeine
done	= (con)done, hydrocodone

40. Oxycodone / Acetaminophen (Percocet)

ox	= ox, (b)ox, sulfamethoxazole/trimethoprim
y	= (countr)y, oxycodone
co'	= co(ne), codeine
done	= (con)done, hydrocodone

a	= (comm)a, (aur)a, acetaminophen
ce	= ce(iling), acetaminophen
ta	= (da)ta, (sona)ta, (quo)ta, acetaminophen
mi'	= mi(nt), mi(lk), acetaminophen
no	= (can)no(n), (ca)no(n), acetaminophen
phen	= (hy)phen, acetaminophen

Opioid analgesics - Schedule III

41. Acetaminophen w/codeine (Tylenol/codeine)

a	= (comm)a, (aur)a, acetaminophen
ce	= ce(iling), acetaminophen
ta	= (da)ta, (sona)ta, (quo)ta, acetaminophen
mi'	= mi(nt), mi(lk), acetaminophen
no	= (can)no(n), (ca)no(n), acetaminophen
phen	= (hy)phen, acetaminophen
co'	= co(ne), codeine
deine	= de(er), (mach)ine, codeine

Mixed-opioid receptor analgesic – Schedule IV

42. Tramadol (Ultram)

tra'	= tra(ck), tramadol
ma	= (paja)ma, somatropin
dol	= dol(l), tramadol

43. Tramadol / Acetaminophen (Ultracet)

tra'	= tra(ck), tramadol
ma	= (paja)ma, somatropin
dol	= dol(l), tramadol
a	= (comm)a, (aur)a, acetaminophen
ce	= ce(iling), acetaminophen
ta	= (da)ta, (sona)ta, (quo)ta, acetaminophen
mi'	= mi(nt), mi(lk), acetaminophen
no	= (can)no(n), (ca)no(n), acetaminophen
phen	= (hy)phen, acetaminophen

Opioid antagonist

44. Naloxone (Narcan)

na	= (bana)na, fexofenadine
lox'	= lox, meloxicam
one	= (c)one, naloxone

45. Buprenorphine / Naloxone (Suboxone) [CIII]

bu	= bu(reau), ibuprofen
pre	= pre(dator), pre(dict), buprenorphine / naloxone
nor'	= nor, nor(th), buprenorphine / naloxone
phine	= (So)phi(a)(du)ne, morphine
na	= (bana)na, fexofenadine
lox'	= lox, meloxicam
one	= (c)one, naloxone

III. HEADACHES AND MIGRAINE

5-HT$_1$ receptor agonist

46. Eletriptan (Relpax)

e	= (n)e(t), etodolac
le	= le(tter), celecoxib
trip'	= trip, trip(lets), eletriptan
tan	= (sul)tan, eletriptan

47. Sumatriptan (Imitrex)

su	= su(e), Su(san), sumatriptan
ma	= (paja)ma, somatropin
trip'	= trip, trip(lets), eletriptan
tan	= (sul)tan, eletriptan

IV. DMARDs AND RHEUMATOID ARTHRITIS

48. Methotrexate (Rheumatrex)

meth	= meth(od), simethicone

o   = "o", fexofenadine
trex'   = tre(mor), "x", methotrexate
ate   = (g)ate, (l)ate, diphenoxylate / atropine

49. Abatacept (Orencia)

a   = (comm)a, (aur)a, acetaminophen
ba   = ba(t), bacitracin
ta'   = (da)ta, (sona)ta, (quo)ta, acetaminophen
cept   = (ac)cept, abatacept

50. Etanercept (Enbrel)

e   = (en)e(my), erythromycin
ta'   = (da)ta, (sona)ta, (quo)ta, acetaminophen
ner   = ner(d), etanercept
cept   = (ac)cept, abatacept

V. OSTEOPOROSIS

Bisphosphonates

51. Alendronate (Fosamax)

a   = (comm)a, (aur)a, acetaminophen
len'   = len(s), len(d), alendronate
dro   = dro(ne), hydrocortisone
nate   = (in)nate, Nate, calcium carbonate

52. Ibandronate (Boniva)

i   = "i", ibuprofen
ban'   = ban(d), ban(ner), ibandronate
dro   = dro(ne), hydrocortisone
nate   = (in)nate, Nate, calcium carbonate

53. Risedronate (Actonel)

ri   = ri(nse), (Flo)ri(da), cetirizine
se'   = se(t), se(nt), ondansetron
dro   = dro(ne), hydrocortisone
nate   = (in)nate, Nate, calcium carbonate

VI. SELECTIVE ESTROGEN RECEPTOR MODULATOR (SERM)

54. Raloxifene (Evista)

ra	= (au)ra, (cob)ra, (Op)ra(h), loperamide
lox'	= lox, meloxicam
i	= i(t), simethicone
fene	= fe(et)(du)ne, raloxifene

VII. MUSCLE RELAXANTS

55. Baclofen (Lioresal)

ba'	= ba(t), bacitracin
clo	= clo(ver), clo(the), diclofenac
fen	= fen, fen(der), ibuprofen

56. Carisoprodol (Soma)

ca	= (or)ca, fluticasone
ri	= ri(nse), (Flo)ri(da), cetirizine
so	= so(n), prednisolone
pro'	= pro(ton), pro(be), pro(active), ibuprofen
dol	= dol(l), tramadol

57. Cyclobenzaprine (Flexeril)

cy	= cy(cle), cy(clone), cyclobenzaprine
clo	= clo(ver), clo(the), diclofenac
ben'	= ben(d), benzocaine
za	= (pla)za, influenza vaccine
prine	= pr(oton), (mach)ine, cyclobenzaprine

58. Diazepam (Valium)

di	= di(ce), diphenhydramine
a'	= (m)a(t), (b)a(t), diphenoxylate / atropine
ze	= ze(d), ze(phyr), azelastine
pam	= pam(phlet), (s)pam, diazepam

Alternate pronunciation:

di	= di(ce), diphenhydramine

a'	= "a", diazepam
ze	= ze(d), ze(phyr), diazepam
pam	= pam(phlet), (s)pam, diazepam

59. Metaxalone (Skelaxin)

me	= me(n), me(sh), esomeprazole
tax'	= tax, tax(i), metaxalone
a	= (comm)a, (aur)a, acetaminophen
lone	= lone, (a)lone, (c)lone, triamcinolone

60. Methocarbamol (Robaxin)

meth	= meth(od), simethicone
o	= "o", fexofenadine
car'	= car, car(d), calcium carbonate
ba	= tu(ba), methocarbamol
mol	= mol(lusk), (enty)mol(ogy), methocarbamol

61. Tizanidine (Zanaflex)

ti	= ti(c), (plas)ti(c), cetirizine
za'	= (pla)za, influenza vaccine
ni	= (k)ni(t), prednisolone
dine	= (bir)di(e)(tu)ne, fexofenadine

Alternate pronunciation:

ti	= ti(e), ti(de), tizanidine
za'	= (pla)za, influenza vaccine
ni	= (k)ni(t), prednisolone
dine	= (bir)di(e)(tu)ne, fexofenadine

VIII. GOUT

62. Colchicine (Colcrys)

col'	= col(lar), Col(orado), polyethylene glycol
chi	= chi(n), chi(p), colchicine
cine	= (leu)cine, influenza vaccine

Uric acid reducers

63. Allopurinol (Zyloprim)

al	= (s)al(iva), Al(exander), albuterol
lo	= (Co)lo(rado), (s)lo(w), loperamide
pur'	= pur(e), pur(ity), allopurinol
i	= i(t), simethicone
nol	= (etha)nol, (mo)nol(ith), allopurinol

64. Febuxostat (Uloric)

fe	= fe(n), fexofenadine
bux'	= bux(om), febuxostat
o	= "o", fexofenadine
stat	= (photo)stat, stat(im), febuxostat

Chapter 3 - Respiratory

Medications

I. Antihistamines and Decongestants

Antihistamine – 1st-generation

65. Diphenhydramine (Benadryl)

di	= di(ce), diphenhydramine
phen	= (hy)phen, acetaminophen
hy'	= hy(brid), hy(phen), magnesium hydroxide
dra	= (hy)dra, diphenhydramine
mine	= (hista)mine, brompheniramine

66. Hydroxyzine (Atarax)

hy	= hy(brid), hy(phen), magnesium hydroxide
drox'	= dr(ink)(b)ox, dr(op)(b)ox, magnesium hydroxide
y	= (t)y(pical), (h)y(pnotize), hydroxyzine
zine	= (maga)zine, cetirizine

OTC Antihistamine – 2nd-generation

67. Cetirizine (Zyrtec)

ce	= ce(nt), ce(nter), cetirizine
ti'	= ti(c), (plas)ti(c), cetirizine
ri	= ri(nse), (Flo)ri(da), cetirizine
zine	= (maga)zine, cetirizine

68. Loratadine (Claritin)

lo	= (Co)lo(rado), (s)lo(w), loperamide
ra'	= ra(ttle), loratadine
ta	= (da)ta, (sona)ta, (quo)ta, acetaminophen
dine	= (bir)di(e)(tu)ne, fexofenadine

OTC Antihistamine – 3nd generation

69. Fexofenadine (Allegra)

fex	= (ponti)fex, fexofenadine
o	= "o", fexofenadine
fe'	= fe(n), fexofenadine
na	= (bana)na, fexofenadine
dine	= (bir)di(e)(tu)ne, fexofenadine

70. Levocetirizine (Xyzal)

le	= le(tter), celecoxib
vo	= vo(te), sevoflurane
ce	= ce(nt), ce(nter), cetirizine
ti'	= ti(c), (plas)ti(c), cetirizine
ri	= ri(nse), (Flo)ri(da), cetirizine
zine	= (maga)zine, cetirizine

OTC Antihistamine – Eye Drops

71. Olopatadine (Patanol, Pataday)

o	= "o", fexofenadine
lo	= (Co)lo(rado), (s)lo(w), loperamide
pa'	= pa(th), palivizumab
ta	= (da)ta, (sona)ta, (quo)ta, acetaminophen
dine	= (bir)di(e)(tu)ne, fexofenadine

Antihistamine – Nasal Spray

72. Azelastine (Astelin)

a	= (m)a(t), (b)a(t), diphenoxylate / atropine
ze'	= ze(d), ze(phyr), azelastine
la	= la(b), la(st), azelastine
stine	= (cele)stine, (cy)stine, azelastine

OTC Antihistamine – 2nd generation / Decongestant

73. Loratadine-D (Claritin-D)

lo	= (Co)lo(rado), (s)lo(w), loperamide

ra'   = ra(ttle), loratadine
ta   = (da)ta, (sona)ta, (quo)ta, acetaminophen
dine   = (bir)di(e)(tu)ne, fexofenadine

pseu   = pseu(do), pseudoephedrine
do   = do(ugh), lidocaine
e   = (n)e(t), etodolac
phe'   = phe(nomenon), brompheniramine
drine   = (alexan)drine, pseudoephedrine

BTC/OTC Decongestants

74. Pseudoephedrine (Sudafed) [BTC]

pseu   = pseu(do), pseudoephedrine
do   = do(ugh), lidocaine
e   = (n)e(t), etodolac
phe'   = phe(nomenon), brompheniramine
drine   = (alexan)drine, pseudoephedrine

75. Phenylephrine (NeoSynephrine) [OTC]

phen   = (hy)phen, acetaminophen
yl   = (s)yl(lable), polyethylene glycol
eph'   = (z)eph(yr), phenylephrine
rine   = (doct)rine, (u)rine, phenylephrine

76. Oxymetazoline (Afrin) [OTC]

ox   = ox, (b)ox, sulfamethoxazole/trimethoprim
y   = (countr)y, oxycodone
me   = me(n), me(sh), esomeprazole
ta'   = ta(b), ta(x), Oxymetazoline
zo   = zo(ne), (o)zo(ne), (gon)zo, benzocaine
line   = line(n), (alka)line [Note: not (fe)line or (sa)line]

II. ALLERGIC RHINITIS STEROID, ANTITUSSIVES, & EXPECTORANTS

Allergic rhinitis steroid

77. Mometasone nasal inhaler (Nasonex)

mo	= mo(at), famotidine
me'	= me(n), me(sh), esomeprazole
ta	= (da)ta, (sona)ta, (quo)ta, acetaminophen
sone	= so(fa), (du)ne, fluticasone

78. Triamcinolone (Nasacort Allergy 24HR)

tri	= tri(angle), triamcinolone
am	= (l)am(p), triamcinolone
ci'	= (s)ci(ntillate), triamcinolone
no	= (can)no(n), (ca)no(n), acetaminophen
lone	= lone, (a)lone, (c)lone, triamcinolone

OTC Antitussive / Expectorant

79. Guaifenesin/DM (Mucinex DM, Robitussin DM)

guai	= gua(va)"i", gua(camole)"i", guaifenesin
fe'	= fe(n), fexofenadine
ne	= ne(t), guaifenesin
sin	= (ba)sin, sin, guaifenesin

dex	= dex(terity), Dex(ter), dextromethorphan
tro	= (me)tro, (re)tro, dextromethorphan
meth	= meth(od), simethicone
or'	= or, (st)or(e), dextromethorphan
phan	= (or)phan, dextromethorphan

RX Antitussive / Expectorant

80. Guaifenesin / Codeine (Cheratussin AC)

guai	= gua(va)"i", gua(camole)"i", guaifenesin
fe'	= fe(n), fexofenadine
ne	= ne(t), guaifenesin
sin	= (ba)sin, sin, guaifenesin

| co | = co(ne), codeine |
| deine | = de(er), (mach)ine, codeine |

RX Antitussive

81. Benzonatate (Tessalon Perles)

ben	= ben(d), benzocaine
zo'	= zo(mbie), benzonatate
na	= (bana)na, fexofenadine
tate	= (es)tate, (mu)tate, benzonatate

III. ASTHMA

Oral steroids

82. Dexamethasone (Decadron)

dex	= dex(terity), Dex(ter), dextromethorphan
a	= (comm)a, (aur)a, acetaminophen
meth'	= meth(od), simethicone
a	= (comm)a, (aur)a, acetaminophen
sone	= so(fa), (du)ne, fluticasone

83. Methylprednisolone (Medrol)

meth	= meth(od), simethicone
yl	= (s)yl(lable), polyethylene glycol
pred	= pred(ator), pred(ict), prednisonolone
ni'	= (k)ni(t), prednisolone
so	= so(n), prednisolone
lone	= lone, (a)lone, (c)lone, triamcinolone

84. Prednisone (Deltasone)

pred'	= pred(ator), pred(ict), prednisonolone
ni	= (k)ni(t), prednisolone
sone	= so(fa), (du)ne, fluticasone

Ophthalmic steroid

85. Loteprednol ophthalmic (Lotemax)

lo	= (Co)lo(rado), (s)lo(w), loperamide
te	= te(nt), motelukast
pred'	= pred(ator), pred(ict), prednisonolone
nol	= (etha)nol, (mo)nol(ith), allopurinol

Inhaled steroid / Beta$_2$ receptor agonist

86. Budesonide / Formoterol (Symbicort)

bu	= bu(reau), ibuprofen
de'	= de(sk), de(n), budesonide
so	= so(n), prednisolone
nide	= (s)nide, (cya)nide, budesonide
for	= for, for(t), formoterol
mo'	= mo(at), famotidine
ter	= ter(se), albuterol
ol	= (aw)ol, (alcoh)ol, albuterol

87. Fluticasone / Salmeterol (Advair)

flu	= flu(id), fluticasone
ti'	= ti(c), (plas)ti(c), cetirizine
ca	= (or)ca, fluticasone
sone	= so(fa), (du)ne, fluticasone
sal	= sal(ad), sal(iva), salmeterol
me'	= me(n), me(sh), esomeprazole
ter	= ter(se), albuterol
ol	= (aw)ol, (alcoh)ol, albuterol

Inhaled steroid

88. Budesonide (Rhinocort, Pulmicort Flexhaler)

bu	= bu(reau), ibuprofen
de'	= de(sk), de(n), budesonide
so	= so(n), prednisolone

nide   = (s)nide, (cya)nide, budesonide

89. Fluticasone (Flonase, Flovent HFA, Flovent Diskus)

flu   = flu(id), fluticasone
ti'   = ti(c), (plas)ti(c), cetirizine
ca   = (or)ca, fluticasone
sone   = so(fa), (du)ne, fluticasone

Beta₂ receptor agonist short acting

90. Albuterol (ProAir HFA, Proventil)

al   = (s)al(iva), Al(exander), albuterol
bu'   = bu(reau), ibuprofen
ter   = ter(se), albuterol
ol   = (aw)ol, (alcoh)ol, albuterol

91. Levalbuterol (Xopenex HFA)

lev   = lev(el), (e)lev(ator), levalbuterol
al   = (s)al(iva), Al(exander), albuterol
bu'   = bu(reau), ibuprofen
ter   = ter(se), albuterol
ol   = (aw)ol, (alcoh)ol, albuterol

Beta₂ receptor agonist / Anticholinergic

92. Albuterol/Ipratropium (DuoNeb)

al   = (s)al(iva), Al(exander), albuterol
bu'   = bu(reau), ibuprofen
ter   = ter(se), albuterol
ol   = (aw)ol, (alcoh)ol, albuterol

ip   = (s)ip, ipratropium
ra   = (au)ra, (cob)ra, (Op)ra(h), loperamide
tro'   = (me)tro, (re)tro, dextromethorphan
pi   = pi(ece), ipratropium
um   = (g)um, calcium carbonate

93. Albuterol / Ipra<u>tropium</u> (Combivent)

al	= (s)al(iva), Al(exander), albuterol
bu'	= bu(reau), ibuprofen
ter	= ter(se), albuterol
ol	= (aw)ol, (alcoh)ol, albuterol
ip	= (s)ip, ipratropium
ra	= (au)ra, (cob)ra, (Op)ra(h), loperamide
tro'	= (me)tro, (re)tro, dextromethorphan
pi	= pi(ece), ipratropium
um	= (g)um, calcium carbonate

Anticholinergic

94. Tio<u>tropium</u> (Spiriva)

ti	= ti(e), ti(de), tizanidine
o	= "o", fexofenadine
tro'	= (me)tro, (re)tro, dextromethorphan
pi	= pi(ece), ipratropium
um	= (g)um, calcium carbonate

Leukotriene receptor antagonist

95. Montelukast (Singulair)

mon	= mon(key), montelukast
te	= te(nt), motelukast
lu'	= lu(cid), montelukast
kast	= (Out)kast, (Di)kast, montelukast

Anti-IgE antibody

96. Oma<u>lizumab</u> (Xolair)

o	= "o", fexofenadine
ma	= (paja)ma, somatropin
li'	= li(ly), li(nt), palivizumab
zu	= Zu(lu), (shiat)zu, palivizumab
mab	= m(at)(c)ab, palivizumab

IV. ANAPHYLAXIS

97. Epinephrine (EpiPen)

e	= (n)e(t), etodolac
pi	= pi(n), aspirin
ne'	= ne(t), guaifenesin
phrine	= phr(ase)(femin)ine, epinephrine

Chapter 4 - Immune

Medications

I. OTC Antimicrobials

Antibiotic cream

98. Neomycin / Polymyxin B / Bacitracin (Neosporin)

ba	= ba(t), bacitracin
ci	= (s)ci(ntillate), triamcinolone
tra'	= tra(y), bacitracin
cin	= cin(der), (s)cin(tillate), bacitracin
ne	= (mo)ne(y), k(ne)e, magnesium hydroxide
o	= "o", fexofenadine
my'	= my, my(osin), neomycin sulfate
cin	= cin(der), (s)cin(tillate), bacitracin
sul'	= sul(len), neomycin sulfate
fate	= fate, neomycin sulfate
po	= po(t), po(lygon), polyethylene glycol
ly	= (li)ly, (po)ly(gon), polyethylene glycol
myx'	= myx(edema), polymyxin B
in	= in, (t)in, polymyxin B
B	= "B"

99. Mupirocin (Bactroban) [RX]

mu	= mu(sic), mu(seum), mupirocin
pi'	= pi(ece), ipratropium
ro	= ro(w), ro(ad), mupirocin
cin	= cin(der), (s)cin(tillate), bacitracin

Antifungal cream

100. Butenafine (Lotrimin Ultra)

bu	= bu(reau), ibuprofen
te'	= te(nt), motelukast
na	= (bana)na, fexofenadine
fine	= fi(eld)(li)ne, butenafine

101. Terbinafine (Lamisil)

ter	= ter(se), albuterol
bi'	= bi(t), (or)bi(t), lubiprostone
na	= (bana)na, fexofenadine
fine	= fi(eld)(li)ne, butenafine

102. Clotrimazole / Betamethasone (Lotrisone) [RX]

clo	= clo(ver), clo(the), diclofenac
tri'	= tri(p), tri(ck), clotrimazole
ma	= (paja)ma, somatropin
zole	= (fe)z(p)ole, sulfamethoxazole/trimethoprim

be	= be(ta), (o)be(y), betamethasone
ta	= (da)ta, (sona)ta, (quo)ta, acetaminophen
meth'	= meth(od), simethicone
a	= (comm)a, (aur)a, acetaminophen
sone	= so(fa), (du)ne, fluticasone

Vaccinations [Some RX, some antibacterial]

These next medications are made up of regular words from biology and microbiology, so I did not include most in the syllable building process.

103. Diphtheria toxoid (Boostrix)

104. Haemophilus influenzae Type B (Pedvax HIB)

105. Influenza vaccine (Fluzone)

in	= in, (t)in, polymyxin B
flu	= flu(id), fluticasone

en'	= (p)en, (m)en, influenza vaccine
za	= (pla)za, influenza vaccine
vac'	= vac(uum), influenza vaccine
cine	= (leu)cine, influenza vaccine

106. Measles, Mumps and Rubella (MMR)

107. Meningococcal (conjugate and polysaccharide) (Menomune)

108. Pertussis in combination

109. Pneumococcal (conjugate and polysaccharide) (Prevnar 13, Pneumovax 23)

110. Polio

111. Rotavirus (RotaTeq)

112. Tetanus in combination

113. Varicella (Varivax)

114. Zoster (Zostavax)

Antiviral OTC

115. Docosanol (Abreva)

do	= do(ugh), lidocaine
co'	= co(ne), codeine
so	= so(n), prednisolone
nol	= (etha)nol, (mo)nol(ith), allopurinol

II. ANTIBIOTICS THAT AFFECT THE CELL WALL

Penicillins

116. Amoxicillin (Amoxil)

a	= (comm)a, (aur)a, acetaminophen

82

mox   = mox(ie), amoxicillin
i   = i(t), simethicone
cil'   = (pen)cil, amoxicillin
lin   = lin(t), amoxicillin

117. Penicillin (Veetids)

pe   = pe(n), pe(ril), pe(t), loperamide
ni   = (k)ni(t), prednisolone
cil'   = (pen)cil, amoxicillin
lin   = lin(t), amoxicillin

Penicillin/Beta-lactamase inhibitor

118. Amoxicillin / Clavulanate (Augmentin)

a   = (comm)a, (aur)a, acetaminophen
mox   = mox(ie), amoxicillin
i   = i(t), simethicone
cil'   = (pen)cil, amoxicillin
lin   = lin(t), amoxicillin

cla'   = cla(w), clavulanate
vu   = (re)vu(e), clavulanate
la   = la(wn), la(w), clavulanate
nate   = (in)nate, Nate, calcium carbonate

Cephalosporins [by generation]

119. Cephalexin (Keflex) [1st]

ce   = ce(nt), ce(nter), cetirizine
pha   = (al)pha, cephalexin
lex'   = (f)lex, cephalexin
in   = in, (t)in, polymyxin B

120. Cefuroxime (Ceftin) [2nd]

cef   = c(l)ef, c(l)ef(t), cefdinir
u'   = "u", cefuroxime
rox   = (xe)rox, (p)rox(imal), cefuroxime
ime   = (ox)ime, cefuroxime

121. C<u>ef</u>dinir (Omnicef) [3rd]

cef'	= c(l)ef, c(l)ef(t), cefdinir
di	= di(nner), di(n), cefdinir
nir	= (souve)nir, cefdinir

122. C<u>ef</u>triaxone (Rocephin) [3rd]

cef	= c(l)ef, c(l)ef(t), cefdinir
tri	= tri(angle), triamcinolone
ax'	= ax, (t)ax(i), ceftriaxone
one	= (c)one, naloxone

123. C<u>ef</u>epime (Maxipime) [4th]

cef'	= c(l)ef, c(l)ef(t), cefdinir
e	= (en)e(my), erythromycin
pime	= pi(ece)(li)me, cefepime

Glycopeptide

124. Vanc<u>omy</u>cin (Vancocin)

van	= van, van(dal), vancomycin
co	= co(ne), codeine
my'	= my, my(osin), neomycin sulfate
cin	= cin(der), (s)cin(tillate), bacitracin

III. ANTIBIOTICS – PROTEIN SYNTHESIS INHIBITORS (BACTERIOSTATIC)

Tetracyclines

125. Doxy<u>cy</u>cline (Doryx)

dox	= (para)dox, (ortho)dox, doxycycline
y	= (countr)y, oxycodone
cy'	= cy(cle), cy(clone), cyclobenzaprine
cline	= cli(ff)(vi)ne, doxycycline

126. Mino<u>cy</u>cline (Minocin)

mi	= mi(nt), mi(lk), acetaminophen

no	= (can)no(n), (ca)no(n), acetaminophen
cy'	= cy(cle), cy(clone), cyclobenzaprine
cline	= cli(ff)(vi)ne, doxycycline

127. Tetracycline (Sumycin)

te	= te(nt), motelukast
tra	= tra(y), bacitracin
cy'	= cy(cle), cy(clone), cyclobenzaprine
cline	= cli(ff)(vi)ne, doxycycline

Macrolides

128. Azithromycin (Zithromax)

a	= (comm)a, (aur)a, acetaminophen
zi	= zi(pper), zi(t), azithromycin
thro	= thro(w), azithromycin
my'	= my, my(osin), neomycin sulfate
cin	= cin(der), (s)cin(tillate), bacitracin

129. Clarithromycin (Biaxin)

cla	= cla(rinet), cla(ret), clarithromycin
ri	= ri(nse), (Flo)ri(da), cetirizine
thro	= thro(w), azithromycin
my'	= my, my(osin), neomycin sulfate
cin	= cin(der), (s)cin(tillate), bacitracin

130. Erythromycin (E-Mycin)

e	= (en)e(my), erythromycin
ry	= (o)ry(x), (t)ry(st), erythromycin
thro	= thro(w), azithromycin
my'	= my, my(osin), neomycin sulfate
cin	= cin(der), (s)cin(tillate), bacitracin

131. Fidaxomicin (Dificid)

fi	= fi(n), fi(b), fidaxomicin
dax'	= (a)dax(ial), (ad)dax, fidaxomicin
o	= "o", fexofenadine
mi'	= mi(le), gentamicin

cin   = cin(der), (s)cin(tillate), bacitracin

Lincosamide

132. Clindamycin (Cleocin)

clin	= clin(ic), clindamycin
da	= (so)da, clindamycin
my'	= my, my(osin), neomycin sulfate
cin	= cin(der), (s)cin(tillate), bacitracin

Oxazolidinone

133. Linezolid (Zyvox)

li	= li(ly), li(nt), palivizumab
ne'	= ne(t), guaifenesin
zo	= (hori)zo(n), trazodone
lid	= lid, slid, linezolid

IV. ANTIBIOTICS – PROTEIN SYNTH. INHIBITORS (BACTERICIDAL)

Aminoglycosides

134. Amikacin (Amikin)

a	= (m)a(t), (b)a(t), diphenoxylate / atropine
mi	= mi(le), fidaxomicin
ka'	= (s)ka(te), Ka(y), amikacin
cin	= cin(der), (s)cin(tillate), bacitracin

135. Gentamicin (Garamycin)

gen	= gen(tleman), gentamicin
ta	= (da)ta, (sona)ta, (quo)ta, acetaminophen
mi'	= mi(le), gentamicin
cin	= cin(der), (s)cin(tillate), bacitracin

V. Antibiotics for urinary tract infections (UTIs) and peptic ulcer disease (PUD)

OTC Urinary tract analgesic

136. Phenazopyridine (Uristat)

phe	= phe(nomenon), brompheniramine
na'	= na(p), (g)na(t), clonazepam
zo	= zo(ne), (o)zo(ne), (gon)zo, benzocaine
py'	= py(lon), py(re), pyrazinamide (PZA)
ri	= ri(nse), (Flo)ri(da), cetirizine
dine	= (bir)di(e)(tu)ne, fexofenadine

Nitrofuran

137. Nitrofurantoin (Macrobid, Macrodantin)

ni	= ni(ght), brompheniramine
tro	= (me)tro, (re)tro, dextromethorphan
fur'	= fur(y), nitrofurantoin
an	= an, an(t), nitrofurantoin
to	= to(e), to(w), pantoprazole
in	= in, (t)in, polymyxin B

Dihydrofolate reductase inhibitor

138. Sulfamethoxazole / Trimethoprim (Bactrim DS)

sul	= sul(len), neomycin sulfate
fa	= (so)fa, sulfamethoxazole/trimethoprim
meth	= meth(od), simethicone
ox'	= ox, (b)ox, sulfamethoxazole/trimethoprim
a	= (comm)a, (aur)a, acetaminophen
zole	= (fe)z(p)ole, sulfamethoxazole/trimethoprim

tri	= tri(angle), triamcinolone
meth'	= meth(od), simethicone
o	= "o", fexofenadine
prim	= prim, prim(p), sulfamethoxazole/trimethoprim

Fluoroquinolones

139. Ciprofloxacin (Cipro)

ci	= (s)ci(ntillate), triamcinolone
pro	= pro(ton), pro(be), pro(active), ibuprofen
flox'	= fl(ocks)ox(en), ciprofloxacin
a	= (comm)a, (aur)a, acetaminophen
cin	= cin(der), (s)cin(tillate), bacitracin

140. Gatifloxacin ophthalmic (Zymar)

ga	= ga(s), ga(p), gatifloxacin
ti	= ti(c), (plas)ti(c), cetirizine
flox'	= fl(ocks)ox(en), ciprofloxacin
a	= (comm)a, (aur)a, acetaminophen
cin	= cin(der), (s)cin(tillate), bacitracin

141. Levofloxacin (Levaquin)

le	= le(tter), celecoxib
vo	= vo(te), sevoflurane
flox'	= fl(ocks)ox(en), ciprofloxacin
a	= (comm)a, (aur)a, acetaminophen
cin	= cin(der), (s)cin(tillate), bacitracin

142. Moxifloxacin (Avelox) / [Ophth. is Vigamox]

mox	= mox(ie), amoxicillin
i	= i(t), simethicone
flox'	= fl(ocks)ox(en), ciprofloxacin
a	= (comm)a, (aur)a, acetaminophen
cin	= cin(der), (s)cin(tillate), bacitracin

Nitroimidazole

143. Metronidazole (Flagyl)

me	= me(n), me(sh), esomeprazole
tro	= (me)tro, (re)tro, dextromethorphan
ni'	= (k)ni(t), prednisolone
da	= (so)da, clindamycin
zole	= (fe)z(p)ole, sulfamethoxazole/trimethoprim

Alternate pronunciation:

me	= me(n), me(sh), esomeprazole
tro	= (me)tro, (re)tro, dextromethorphan
ni'	= ni(ght), brompheniramine
da	= (so)da, clindamycin
zole	= (fe)z(p)ole, sulfamethoxazole/trimethoprim

VI. ANTI-TUBERCULOSIS AGENTS

144. Rifampin (Rifadin)

ri	= ri(nse), (Flo)ri(da), cetirizine
fam'	= fam(ily), fam(ine), rifampin
pin	= pin, somatropin

145. Isoniazid (INH)

i	= "i", ibuprofen
so	= so(n), prednisolone
ni'	= ni(ght), brompheniramine
a	= (comm)a, (aur)a, acetaminophen
zid	= zi(ppe)d, isoniazid(INH)

146. Pyrazinamide (PZA)

py	= py(lon), py(re), pyrazinamide (PZA)
ra	= (au)ra, (cob)ra, (Op)ra(h), loperamide
zi'	= zi(pper), zi(t), azithromycin
na	= (bana)na, fexofenadine
mide	= (gu)m(sl)ide, loperamide

147. Ethambutol (Myambutol)

eth	= (m)eth(od), polyethylene glycol
am'	= (I)am(p), triamcinolone
bu	= bu(reau), ibuprofen
tol	= tol(erate), (derma)tol(ogy), ethambutol

VII. ANTIFUNGALS

148. Amphotericin B (Fungizone)

am	= (l)am(p), triamcinolone
pho	= pho(ne), pho(to), amphotericin B
te'	= te(nt), motelukast
ri	= ri(nse), (Flo)ri(da), cetirizine
cin	= cin(der), (s)cin(tillate), bacitracin

149. Fluconazole (Diflucan)

flu	= flu(id), fluticasone
co'	= co(ne), codeine
na	= (bana)na, fexofenadine
zole	= (fe)z(p)ole, sulfamethoxazole/trimethoprim

150. Ketoconazole (Nizoral)

ke	= ke(y), ke(ep), ketoconazole
to	= (pis)to(n), (a)to(m), ketoconazole
co'	= co(ne), codeine
na	= (bana)na, fexofenadine
zole	= (fe)z(p)ole, sulfamethoxazole/trimethoprim

151. Nystatin (Mycostatin)

ny'	= ny(lon), nystatin
sta	= sta(ff), nystatin
tin	= tin, tin(der), nystatin

VIII. ANTIVIRALS – NON-HIV

Influenza A and B

152. Oseltamivir (Tamiflu)

o	= "o", fexofenadine
sel	= sel(fie), sel(l), oseltamivir
ta'	= (da)ta, (sona)ta, (quo)ta, acetaminophen
mi	= mi(nt), mi(lk), acetaminophen
vir	= vir(ulent), oseltamivir [rhymes with veer]

153. Zanamivir (Relenza)

za	= (pla)za, influenza vaccine
na'	= (bana)na, fexofenadine
mi	= mi(nt), mi(lk), acetaminophen
vir	= vir(ulent), oseltamivir [rhymes with veer]

Herpes simplex virus & Varicella-Zoster Virus HSV/VSV

154. Acyclovir (Zovirax)

a	= (comm)a, (aur)a, acetaminophen
cy'	= cy(cle), cy(clone), cyclobenzaprine
clo	= clo(ver), clo(the), diclofenac
vir	= vir(ulent), oseltamivir [rhymes with veer]

155. Valacyclovir (Valtrex)

val	= val(ley), val(ve), divalproex
a	= (comm)a, (aur)a, acetaminophen
cy'	= cy(cle), cy(clone), cyclobenzaprine
clo	= clo(ver), clo(the), diclofenac
vir	= vir(ulent), oseltamivir [rhymes with veer]

Respiratory Syncytial Virus RSV

156. Palivizumab (Synagis)

pa	= pa(th), palivizumab
li	= li(ly), li(nt), palivizumab
vi'	= vi(sion), vi(m), palivizumab
zu	= Zu(lu), (shiat)zu, palivizumab
mab	= m(at)(c)ab, palivizumab

Hepatitis

157. Entecavir (Baraclude)

en	= (p)en, (m)en, influenza vaccine
te'	= te(nt), motelukast
ca	= (or)ca, fluticasone
vir	= vir(ulent), oseltamivir [rhymes with veer]

158. Hepatitis A (Havrix)

159. Hepatitis B (Recombivax HB)

HPV

160. Human papillomavirus (Gardasil)

IX. ANTIVIRALS – HIV

Fusion Inhibitor

161. Enfuvirtide (Fuzeon) (T-20)

en	= (p)en, (m)en, influenza vaccine
fu'	= fu(mes), fu(ry), enfuvirtide (T-20)
vir	= vir(tue), enfuvirtide
tide	= tide, (rip)tide, enfuvirtide (T-20)

CCR5 Antagonist

162. Maraviroc (Selzentry) (MVC)

Non-nucleoside reverse transcriptase inhibitors (NNRTI) with two nucleoside / nucleotide reverse transcriptase inhibitors (NRTIs)

ma	= (paja)ma, somatropin
ra	= ra(ttle), loratadine
vi'	= vi(sion), vi(m), palivizumab
roc	= roc(k), (c)rock, maraviroc

163. Efavirenz (Sustiva) [NNRTI]

e	= (n)e(t), etodolac
fa'	= fa(ct), fa(ctor), efavirenz
vir	= vir(tue), enfuvirtide
enz	= enz(yme), (fr)enz(y), efavirenz

164. Emtricitabine / Tenofovir (Truvada) [NRTIs]

em	= (g)em, em(ber), emtricitabine
tri	= tri(angle), triamcinolone
ci'	= ci(der), (s)ci(ence), emtricitabine

ta	= (da)ta, (sona)ta, (quo)ta, acetaminophen
bine	= (zom)bi(e)(li)ne, (Yohim)bine, emtricitabine
te	= te(nt), motelukast
no'	= no(d), no(t), tenofovir
fo	= (ef)fo(rt), tenofovir
vir	= vir(ulent), oseltamivir [rhymes with veer]

165. Efavirenz / Emtricitabine / Tenofovir (Atripla) [NNRTI / NRTIs] (EFV / FTC / TDF)

e	= (n)e(t), etodolac
fa'	= fa(ct), fa(ctor), efavirenz
vir	= vir(tue), enfuvirtide
enz	= enz(yme), (fr)enz(y), efavirenz
em	= (g)em, em(ber), emtricitabine
tri	= tri(angle), triamcinolone
ci'	= ci(der), (s)ci(ence), emtricitabine
ta	= (da)ta, (sona)ta, (quo)ta, acetaminophen
bine	= (zom)bi(e)(li)ne, (Yohim)bine, emtricitabine
te	= te(nt), motelukast
no'	= no(d), no(t), tenofovir
fo	= (ef)fo(rt), tenofovir
vir	= vir(ulent), oseltamivir [rhymes with veer]

Integrase Strand Transfer Inhibitor

166. Raltegravir (Isentress) (RAL)

ral	= ral(ly), Ral(ph), raltegravir
te'	= te(nt), motelukast
gra	= (ag)gra(vate), raltegravir
vir	= vir(ulent), oseltamivir [rhymes with veer]

Protease Inhibitor

167. Atazanavir (Reyataz) (ATV)

a	= (m)a(t), (b)a(t), diphenoxylate / atropine

ta	= (da)ta, (sona)ta, (quo)ta, acetaminophen
za'	= (pla)za, influenza vaccine
na	= (bana)na, fexofenadine
vir	= vir(ulent), oseltamivir [rhymes with veer]

168. Darunavir (Prezista) (DRV)

da	= (so)da, clindamycin
ru'	= ru(by), ru(in), darunavir
na	= (bana)na, fexofenadine
vir	= vir(ulent), oseltamivir [rhymes with veer]

X. MISCELLANEOUS

169. Albendazole (Albenza) [Anthelmintic]

al	= (s)al(iva), Al(exander), albuterol
ben'	= ben(d), benzocaine
da	= (so)da, clindamycin
zole	= (fe)z(p)ole, sulfamethoxazole/trimethoprim

170. Hydroxychloroquine (Plaquenil)

hy	= hy(brid), hy(phen), magnesium hydroxide
drox	= dr(ink)(b)ox, dr(op)(b)ox, magnesium hydroxide
y	= (countr)y, oxycodone
chlor'	= chlor(ine), chlorpheniramine
o	= "o", fexofenadine
quine	= quin(c)e, quin(t)e(t), hydroxychloroquine

Antimalarial

171. Nitazoxanide (Alinia) [Antiprotozoal]

ni'	= ni(ght), brompheniramine
ta	= (da)ta, (sona)ta, (quo)ta, acetaminophen
zox'	= (qui)z(b)ox, nitazoxanide
a	= (comm)a, (aur)a, acetaminophen
nide	= (s)nide, (cya)nide, budesonide

CHAPTER 5 - NEURO

MEDICATIONS

I. OTC LOCAL ANESTHETICS AND ANTIVERTIGO

Local anesthetics

172. Benzocaine (Anbesol) [Ester type]

ben'	= ben(d), benzocaine
zo	= zo(ne), (o)zo(ne), (gon)zo, benzocaine
caine	= c(at)(migr)aine, benzocaine

173. Lidocaine (Solarcaine) [Amide type]

li'	= li(ght), li(e), lidocaine
do	= do(ugh), lidocaine
caine	= c(at)(migr)aine, benzocaine

Antivertigo

174. Meclizine (Dramamine, Antivert [RX])

me'	= me(n), me(sh), esomeprazole
cli	= cli(ff), meclizine
zine	= (maga)zine, cetirizine

II. SEDATIVE-HYPNOTICS (SLEEPING PILLS)

OTC Non-narcotic analgesic / Sedative-hypnotic

175. Acetaminophen PM (Tylenol PM)

a	= (comm)a, (aur)a, acetaminophen
ce	= ce(iling), acetaminophen
ta	= (da)ta, (sona)ta, (quo)ta, acetaminophen
mi'	= mi(nt), mi(lk), acetaminophen

no   = (can)no(n), (ca)no(n), acetaminophen
phen   = (hy)phen, acetaminophen

di   = di(ce), diphenhydramine
phen   = (hy)phen, acetaminophen
hy'   = hy(brid), hy(phen), magnesium hydroxide
dra   = (hy)dra, diphenhydramine
mine   = (hista)mine, brompheniramine

Benzodiazepine-like

176. Eszopiclone (Lunesta)

es   = (m)es(s), Es(ther), esomeprazole
zo   = zo(ne), (o)zo(ne), (gon)zo, benzocaine
pi'   = pi(n), aspirin
clone   = clone, (cy)clone, eszopiclone

177. Zolpidem (Ambien)

zol'   = zo(na)l, zolpidem
pi   = pi(n), aspirin
dem   = dem(ocracy), dem(and), zolpidem

Melatonin receptor agonist

178. Ramelteon (Rozerem)

ra   = ra(ttle), loratadine
mel'   = mel(on), mel(t), ramelteon
te   = te(a), te(e), ramelteon
on   = on, (w)on(ton), ondansetron

Miscellaneous

179. Trazodone (Desyrel)

tra'   = tra(y), bacitracin
zo   = (hori)zo(n), trazodone
done   = (con)done, hydrocodone

III. ANTIDEPRESSANTS

Miscellaneous / SSRI

180. Vilazodone (Viibryd)

vi	= vi(sion), vi(m), palivizumab
la'	= la(ser), la(ce), vilazodone
zo	= (hori)zo(n), trazodone
done	= (con)done, hydrocodone

Selective serotonin reuptake inhibitors (SSRIs)

181. Citalopram (Celexa)

ci	= (s)ci(ntillate), triamcinolone
ta'	= ta(b), ta(x), Oxymetazoline
lo	= (Co)lo(rado), (s)lo(w), loperamide
pram	= pr(e)am(ble), pram, citalopram

182. Escitalopram (Lexapro)

es	= (m)es(s), Es(ther), esomeprazole
ci	= (s)ci(ntillate), triamcinolone
ta'	= ta(b), ta(x), Oxymetazoline
lo	= (Co)lo(rado), (s)lo(w), loperamide
pram	= pr(e)am(ble), pram, citalopram

183. Sertraline (Zoloft)

ser'	= (dre)ser, ser(f), sertraline
tra	= (orches)tra, (spec)tra, (ex)tra, sertraline
line	= (sa)line, sertraline

184. Fluoxetine (Prozac, Sarafem)

flu	= flu(id), fluticasone
ox'	= ox, (b)ox, sulfamethoxazole/trimethoprim
e	= (n)e(t), etodolac
tine	= (sal)tine, (rou)tine, fluoxetine

185. Paroxetine (Paxil, Paxil CR)

pa	= pa(th), palivizumab

rox'   = (xe)rox, p(rox)imal, cefuroxime
e   = (n)e(t), etodolac
tine   = (sal)tine, (rou)tine, fluoxetine

Serotonin-Norepinephrine reuptake inhibitors (SNRIs)

186. Duloxetine (Cymbalta)

du   = du(et), du(ne), duloxetine
lox'   = lox , meloxicam
e   = (n)e(t), etodolac
tine   = (sal)tine, (rou)tine, fluoxetine

187. Desvenlafaxine (Pristiq)

des   = des(ks), des(tination), desvenlafaxine
ven   = ven(om), ven(dor), desvenlafaxine
la   = (co)la, (Koa)la, desvenlafaxine
fax'   = fax, (tele)fax, desvenlafaxine
ine   = (sal)ine, desvenlafaxine

188. Venlafaxine (Effexor)

ven   = ven(om), ven(dor), desvenlafaxine
la   = (co)la, (Koa)la, desvenlafaxine
fax'   = fax, (tele)fax, desvenlafaxine
ine   = (sal)ine, desvenlafaxine

Tricyclic antidepressants (TCAs)

189. Amitriptyline (Elavil)

a   = (m)a(t), (b)a(t), diphenoxylate / atropine
mi   = mi(nt), mi(lk), acetaminophen
trip'   = trip, trip(lets), eletriptan
ty   = ty(pical), ty(mpanic), amitriptyline
line   = (sa)line, sertraline

190. Doxepin (Sinequan)

dox'   = (para)dox, (ortho)dox, doxycycline
e   = (n)e(t), etodolac

pin   = pin, somatropin

191. Nor<u>triptyline</u> (Pamelor)

nor   = nor, nor(th), buprenorphine / naloxone
trip'   = trip, trip(lets), eletriptan
ty   = ty(pical), ty(mpanic), amitriptyline
line   = (sa)line, sertraline

Tetracycylic antidepressant (TeCA) Noradrenergic and specific serotonergic antidepressants (NaSSAs)

192. Mirtazapine (Remeron)

mir   = (ad)mir(al), mir(th), mirtazapine
ta'   = ta(b), ta(x), Oxymetazoline
za   = (pla)za, influenza vaccine
pine   = (shi)p(mar)ine, atropine

Alternate pronunciation:

mir   = (ad)mir(al), mir(th), mirtazapine
ta'   = ta(pe), ta(me), mirtazapine
za   = (pla)za, influenza vaccine
pine   = (shi)p(mar)ine, atropine

Monoamine oxidase inhibitor (MAOI)

193. Isocarboxazid (Marplan)

i   = "i", ibuprofen
so   = so(fa), so, docusate sodium
car   = car, car(d), calcium carbonate
box'   = box, isocarboxazid
a   = (comm)a, (aur)a, acetaminophen
zid   = zi(ppe)d, isoniazid(INH)

IV. SMOKING CESSATION

194. Bupropion (Wellbutrin, Zyban)

bu   = bu(reau), ibuprofen

pro'	= pro(ton), pro(be), pro(active), ibuprofen
pi	= pi(ece), ipratropium
on	= on, (w)on(ton), ondansetron

195. Varenicline (Chantix)

va	= (lar)va, (sali)va), varenicline
re'	= re(d), re(f), varenicline
ni	= (k)ni(t), prednisolone
cline	= cl(iff)(sal)ine, varenicline

V. BARBITURATES

196. Phenobarbital (Luminal)

phe	= phe(nom), phe(notype), phenobarbital
no	= no, (k)no(w), phenobarbital
bar'	= bar, bar(bell), phenobarbital
bi	= bi(t), (or)bi(t), lubiprostone
tal	= tal(l), butalbital

VI. BENZODIAZEPINES

197. Alprazolam (Xanax)

al	= (s)al(iva), Al(exander), albuterol
pra'	= pra(ctice), pra(ttle), pramipexole
zo	= zo(ne), (o)zo(ne), (gon)zo, benzocaine
lam	= lam(b), lam(p), alprazolam

Alternate pronunciation:

al	= (s)al(iva), Al(exander), albuterol
pra'	= pra(y), pra(ise), alprazolam
zo	= zo(ne), (o)zo(ne), (gon)zo, benzocaine
lam	= lam(b), lam(p), alprazolam

198. Midazolam (Versed)

| mi | = mi(nt), mi(lk), acetaminophen |
| da' | = da(y), da(te), midazolam |

zo   = zo(ne), (o)zo(ne), (gon)zo, benzocaine
lam   = lam(b), lam(p), alprazolam

199. Clonazepam (Klonopin)

clo   = clo(ver), clo(the), diclofenac
na'   = na(vy), na(me), clonazepam
ze   = ze(d), ze(phyr), azelastine
pam   = pam(phlet), (s)pam, diazepam

Alternate pronunciation:

clo   = clo(th), clo(g), clonazepam
na'   = na(p), (g)na(t), clonazepam
ze   = (ha)ze(l), clonazepam
pam   = pam(phlet), (s)pam, diazepam

200. Lorazepam (Ativan)

lo   = (Co)lo(rado), (s)lo(w), loperamide
ra'   = ra(y), ra(zor), lorazepam
ze   = ze(d), ze(phyr), azelastine
pam   = pam(phlet), (s)pam, diazepam

Alternate pronunciation:

lo   = (Co)lo(rado), (s)lo(w), loperamide
ra'   = ra(ttle), loratadine
ze   = (ha)ze(l), clonazepam
pam   = pam(phlet), (s)pam, diazepam

201. Temazepam (Restoril)

te   = te(nt), motelukast
ma'   = ma(ze), may, temazepam
ze   = ze(d), ze(phyr), azelastine
pam   = pam(phlet), (s)pam, diazepam

Alternate pronunciation:

te   = te(nt), motelukast
ma'   = ma(p), ma(t), temazepam
ze   = (ha)ze(l), clonazepam

pam   = pam(phlet), (s)pam, diazepam

VII. Non-Benzodiazepine / Non-Barbiturate

202. Buspirone (Buspar)

bu'   = bu(reau), ibuprofen
spi   = spi(ll), spi(n), buspirone
rone   = (d)rone, (p)rone, buspirone

VIII. ADHD Medications

Stimulant – Schedule II

203. Amphetamine/Dextroamphetamine (Adderall)

am   = (l)am(p), triamcinolone
phe'   = phe(nomenon), brompheniramine
ta   = (da)ta, (sona)ta, (quo)ta, acetaminophen
mine   = (hista)mine, brompheniramine

dex   = dex(terity), Dex(ter), dextromethorphan
tro   = (me)tro, (re)tro, dextromethorphan
am   = (l)am(p), triamcinolone
phe'   = phe(nomenon), brompheniramine
ta   = (da)ta, (sona)ta, (quo)ta, acetaminophen
mine   = (hista)mine, brompheniramine

204. Dexmethylphenidate (Focalin)

dex   = dex(terity), Dex(ter), dextromethorphan
meth   = meth(od), simethicone
yl   = (s)yl(lable), polyethylene glycol
phe'   = phe(nomenon), brompheniramine
ni   = (k)ni(t), prednisolone
date   = date, (up)date, dexmethylphenidate

205. Lisdexamfetamine (Vyvanse)

lis   = lis(t), lis(ten), lisdexamfetamine
dex   = dex(terity), Dex(ter), dextromethorphan

am	= (l)am(p), triamcinolone
fe'	= fe(n), fexofenadine
ta	= (da)ta, (sona)ta, (quo)ta, acetaminophen
mine	= (hista)mine, brompheniramine

206. Methylphenidate (Concerta)

meth	= meth(od), simethicone
yl	= (s)yl(lable), polyethylene glycol
phe'	= phe(nomenon), brompheniramine
ni	= (k)ni(t), prednisolone
date	= date, (up)date, dexmethylphenidate

Non-stimulant – non-scheduled

207. Atomoxetine (Strattera)

a	= "a", diazepam
to	= (pis)to(n), (a)to(m), ketoconazole
mox'	= mox(ie), amoxicillin
e	= (n)e(t), etodolac
tine	= (sal)tine, (rou)tine, fluoxetine

IX. BIPOLAR DISORDER

Simple salt

208. Lithium (Lithobid)

li'	= li(ly), li(nt), palivizumab
thi	= thi(ef), lithium
um	= (g)um, calcium carbonate

X. SCHIZOPHRENIA

First generation antipsychotic (FGA) (low potency)

209. Chlorpromazine (Thorazine)

| chlor | = chlor(ine), chlorpheniramine |
| pro' | = pro(ton), pro(be), pro(active), ibuprofen |

ma	= (paja)ma, somatropin
zine	= (maga)zine, cetirizine

First generation antipsychotic (FGA) (high potency)

210. Haloperidol (Haldol)

ha	= ha(t), ha(lf), haloperidol
lo	= (Co)lo(rado), (s)lo(w), loperamide
pe'	= pe(n), pe(ril), pe(t), loperamide
ri	= ri(nse), (Flo)ri(da), cetirizine
dol	= dol(l), tramadol

Second-generation antipsychotic (SGA)

211. Aripiprazole (Abilify)

a	= a(ir), aripiprazole
ri	= ri(nse), (Flo)ri(da), cetirizine
pi'	= pi(n), aspirin
pra	= (su)pra, (O)pra(h), dexlansoprazole
zole	= (fe)z(p)ole, sulfamethoxazole/trimethoprim

212. Olanzapine (Zyprexa)

o	= "o", fexofenadine
lan'	= lan(ce), lan(d), dexlansoprazole
za	= (pla)za, influenza vaccine
pine	= (shi)p(mar)ine, atropine

213. Risperidone (Risperdal)

ri	= ri(nse), (Flo)ri(da), cetirizine
spe'	= spe(ck), spe(nd), risperidone
ri	= ri(nse), (Flo)ri(da), cetirizine
done	= (con)done, hydrocodone

214. Quetiapine (Seroquel)

que	= que(ll), que(st), quetiapine
ti'	= ti(e), ti(de), tizanidine
a	= (comm)a, (aur)a, acetaminophen
pine	= (shi)p(mar)ine, atropine

XI. ANTIEPILEPTICS

Traditional antiepileptics

215. Carbamazepine (Tegretol)

car	= car, car(d), calcium carbonate
ba	= (tu)ba, methocarbamol
ma'	= ma(ze), may, temazepam
ze	= ze(d), ze(phyr), azelastine
pine	= (shi)p(mar)ine, atropine

216. Divalproex (Depakote)

di	= di(ce), diphenhydramine
val'	= val(ley), val(ve), divalproex
pro	= pro(ton), pro(be), pro(active), ibuprofen
ex	= ex(po), ex(it), divalproex [Note: not "egg zit"]

217. Phenytoin (Dilantin)

phe	= phe(nomenon), brompheniramine
ny'	= (po)ny, (to)ny, phenytoin
to	= to(e), to(w), pantoprazole
in	= in, (t)in, polymyxin B

Newer antiepileptics

218. Gabapentin (Neurontin)

ga'	= ga(s), ga(p), gatifloxacin
ba	= (tu)ba, methocarbamol
pen	= pen, pen(ny), gabapentin
tin	= tin, tin(der), nystatin

219. Lamotrigine (Lamictal)

la	= la(b), la(st), azelastine
mo'	= mo(at), famotidine
tri	= tri(p), tri(ck), clotrimazole
gine	= (aspara)gine, lamotrigine

220. Levetiracetam (Keppra)

le	= le(tter), celecoxib
ve	= ve(teran), ve(st), carvedilol
ti	= ti(ara), (zi)ti, levetiracetam
ra'	= (au)ra, (cob)ra, (Op)ra(h), loperamide
ce	= ce(iling), acetaminophen
tam	= tam(ale), levetiracetam

221. Oxcarbazepine (Trileptal)

ox'	= ox, (b)ox, sulfamethoxazole/trimethoprim
car	= car, car(d), calcium carbonate
ba'	= ba(y), ba(it), oxcarbazepine
ze	= ze(d), ze(phyr), azelastine
pine	= (shi)p(mar)ine, atropine

222. Pregabalin (Lyrica)

pre	= pre(mie), pre(med), pregabalin
ga'	= ga(s), ga(p), gatifloxacin
ba	= (tu)ba, methocarbamol
lin	= lin(t), amoxicillin

223. Topiramate (Topamax)

to	= to(e), to(w), pantoprazole
pi'	= pi(ece), ipratropium
ra	= (au)ra, (cob)ra, (Op)ra(h), loperamide
mate	= mate, (room)mate, topiramate

224. Ziprasidone (Geodon)

zi	= zi(pper), zi(t), azithromycin
pra'	= pra(y), pra(ise), alprazolam
si	= si(t), simethicone
done	= (con)done, hydrocodone

XII. Parkinson's, Alzheimer's, motion sickness

Parkinson's

225. Benz<u>tropine</u> mesylate (Cogentin)

benz	= (Mercedes)Benz, benz(ene), benztropine
tro'	= (me)tro, (re)tro, dextromethorphan
pine	= (shi)p(mar)ine, atropine

226. Levo<u>dopa</u> / Carbi<u>dopa</u> (Sinemet)

le	= le(tter), celecoxib
vo	= vo(te), sevoflurane
do'	= do(ugh), lidocaine
pa	= (pa)pa, levodopa
car	= car, car(d), calcium carbonate
bi	= bi(t), (or)bi(t), lubiprostone
do'	= do(ugh), lidocaine
pa	= (pa)pa, levodopa

227. Sele<u>gi</u>line (Eldepryl)

se	= se(t), se(nt), ondansetron
le'	= le(tter), celecoxib
gi	= (ma)gi(c), (lo)gi(c), selegiline
line	= (sa)line, sertraline

228. Pramipexole (Mirapex ER)

pra	= pra(ctice), pra(ttle), alprazolam
mi	= mi(nt), mi(lk), acetaminophen
pex'	= (a)pex, pramipexole
ole	= (m)ole, (h)ole, pramipexole

229. Ropinirole (Requip, Requip XL)

ro	= ro(w), ro(ad), mupirocin
pi'	= pi(n), aspirin
ni	= (k)ni(t), prednisolone
role	= role, (pa)role, ropinirole

Alzheimer's

230. Donepezil (Aricept)

do	= do(t), do(dge), docusate sodium
ne'	= ne(t), guaifenesin
pe	= pe(n), pe(ril), pe(t), loperamide
zil	= (Bra)zil, zil(ch), donepezil

231. Memantine (Namenda)

me	= me(n), me(sh), esomeprazole
man'	= hu(man), memantine
tine	= (sal)tine, (rou)tine, fluoxetine

Motion sickness

232. Scopolamine (Transderm-Scop)

sco	= sco(pe), sco(ne), scopolamine
po'	= po(t), po(lygon), polyethylene glycol
la	= la(wn), la(w), clavulanate
mine	= (hista)mine, brompheniramine

CHAPTER 6 - CARDIO

MEDICATIONS

I. OTC ANTIHYPERLIPIDEMICS AND ANTIPLATELET

Antihyperlipidemics

233. Omega-3-acid ethyl esters (Lovaza)

234. Niacin (Niaspan ER)

ni'	= ni(ght), brompheniramine
a	= (comm)a, (aur)a, acetaminophen
cin	= cin(der), (s)cin(tillate), bacitracin

Antiplatelet

235. Aspirin (Ecotrin)

as'	= as(p), aspirin
pi	= pi(n), aspirin
rin	= rin(se), permethrin, [Note: not r-i-n-e]

II. DIURETICS

Osmotic

236. Mannitol (Osmitrol)

man'	= man, mannitol
ni	= (k)ni(t), prednisolone
tol	= tol(erate), (derma)tol(ogy), ethambutol

Loop

237. Furosemide (Lasix)

fu	= fu(mes), fu(ry), enfuvirtide (T-20)
ro'	= ro(w), ro(ad), mupirocin
se	= se(t), se(nt), ondansetron
mide	= (gu)m(sl)ide, loperamide

Thiazide

238. Hydrochlorothiazide (Microzide)

hy	= hy(brid), hy(phen), magnesium hydroxide
dro	= dro(ne), hydrocortisone
chlor	= chlor(ine), chlorpheniramine
o	= "o", fexofenadine
thi'	= thi(gh), hydrochlorothiazide
a	= (comm)a, (aur)a, acetaminophen
zide	= z(ipper)(r)ide, hydrochlorothiazide

Potassium sparing and thiazide

239. Triamterene/Hydrochlorothiazide (Dyazide)

tri	= tri(angle), triamcinolone
am'	= (l)am(p), triamcinolone
te	= te(nt), motelukast
rene	= (se)rene, triamterene
hy	= hy(brid), hy(phen), magnesium hydroxide
dro	= dro(ne), hydrocortisone
chlor	= chlor(ine), chlorpheniramine
o	= "o", fexofenadine
thi'	= thi(gh), hydrochlorothiazide
a	= (comm)a, (aur)a, acetaminophen
zide	= z(ipper)(r)ide, hydrochlorothiazide

Potassium sparing

240. Spironolactone (Aldactone)

spi	= spi(ll), spi(n), spironolactone
ro	= ro(w), ro(ad), mupirocin
no	= no, (k)no(w), phenobarbital
lac'	= (li)lac, lac(tose), spironolactone
tone	= tone, (s)tone, nabumetone

Electrolyte replenishment

241. Potassium chloride (K-DUR)

po	= po(tato), propofol
tas'	= tas(k), potassium
si	= (enthu)si(astic), (tran)si(ent), magnesium hydroxide
um	= (g)um, calcium carbonate
chlor'	= chlor(ine), chlorpheniramine
ide	= (sl)ide, magnesium hydroxide

III. UNDERSTANDING THE ALPHAS AND BETAS

Alpha-1 antagonist

242. Doxazosin (Cardura)

dox	= (para)dox, (ortho)dox, doxycycline
a'	= (m)a(t), (b)a(t), diphenoxylate / atropine
zo	= zo(ne), (o)zo(ne), (gon)zo, benzocaine
sin	= (ba)sin, sin, guaifenesin

243. Terazosin (Hytrin)

ter	= ter(se), albuterol
ra'	= ra(ttle), loratadine
zo	= zo(ne), (o)zo(ne), (gon)zo, benzocaine
sin	= (ba)sin, sin, guaifenesin

Alpha-2 agonist

244. Clonidine (Catapres)

clo'	= clo(th), clo(g), clonazepam
ni	= (k)ni(t), prednisolone
dine	= (bir)di(e)(tu)ne, fexofenadine

Beta-blocker – 1st-generation – non-beta selective

245. Propranolol (Inderal)

pro	= pro(ton), pro(be), pro(active), ibuprofen
pra'	= pra(nce), (so)pra(no), propranolol
no	= (can)no(n), (ca)no(n), acetaminophen
lol	= lol(lipop), lol(lygag), propranolol

Beta-blockers – 2nd-generation – beta selective

246. Atenolol (Tenormin)

a	= (comm)a, (aur)a, acetaminophen
te'	= te(nt), motelukast
no	= (can)no(n), (ca)no(n), acetaminophen
lol	= lol(lipop), lol(lygag), propranolol

247. Atenolol / Chlorthalidone (Tenoretic)

a	= (comm)a, (aur)a, acetaminophen
te'	= te(nt), motelukast
no	= (can)no(n), (ca)no(n), acetaminophen
lol	= lol(lipop), lol(lygag), propranolol
chlor	= chlor(ine), chlorpheniramine
tha'	= tha(tch), tha(t), chlorthalidone
li	= li(ly), li(nt), palivizumab
done	= (con)done, hydrocodone

248. Bisoprolol / Hydrochlorothiazide (Ziac)

bi	= bi(t), (or)bi(t), lubiprostone
so'	= so(fa), so, docusate sodium
pro	= (a)pro(n), metoprolol

lol   = lol(lipop), lol(lygag), propranolol

hy   = hy(brid), hy(phen), magnesium hydroxide
dro   = dro(ne), hydrocortisone
chlor   = chlor(ine), chlorpheniramine
o   = "o", fexofenadine
thi'   = thi(gh), hydrochlorothiazide
a   = (comm)a, (aur)a, acetaminophen
zide   = z(ipper)(r)ide, hydrochlorothiazide

249. Metoprolol succinate (Toprol-XL) add salt

me   = me(n), me(sh), esomeprazole
to'   = to(e), to(w), pantoprazole
pro   = (a)pro(n), metoprolol
lol   = lol(lipop), lol(lygag), propranolol

suc'   = suc(ces), suc(tion), metoprolol succinate
ci   = (s)ci(ntillate), triamcinolone
nate   = (in)nate, Nate, calcium carbonate

250. Metoprolol tartrate (Lopressor)

me   = me(n), me(sh), esomeprazole
to'   = to(e), to(w), pantoprazole
pro   = (a)pro(n), metoprolol
lol   = lol(lipop), lol(lygag), propranolol

tar'   = tar, tar(get), metoprolol tartrate
trate   = (concen)trate, (magis)trate, metoprolol tartrate

Beta-blocker – 3rd-generation – non-beta selective vasodilating

251. Carvedilol (Coreg)

car'   = car, car(d), calcium carbonate
ve   = ve(teran), ve(st), carvedilol
di   = di(nner), di(n), cefdinir
lol   = lol(lipop), lol(lygag), propranolol

252. Labet<u>alol</u> (Normodyne)

la	= la(b), la(st), azelastine
be'	= be(ta), o(be)y, betamethasone
ta	= (da)ta, (sona)ta, (quo)ta, acetaminophen
lol	= lol(lipop), lol(lygag), propranolol

253. Nebiv<u>olol</u> (Bystolic)

ne	= ne(t), guaifenesin
bi'	= bi(t), (or)bi(t), lubiprostone
vo	= (di)vo(t), (pi)vo(t), nebivolol
lol	= lol(lipop), lol(lygag), propranolol

IV. Renin-Angiotensin-Aldosterone System (RAAS)

<u>ACE Inhibitors (ACEIs)</u>

254. Benaze<u>pril</u> / HCTZ (Lotensin HCT)

be	= be(d), rabeprazole
na'	= na(p), (g)na(t), clonazepam
ze	= ze(d), ze(phyr), azelastine
pril	= pril(l), benazepril

hy	= hy(brid), hy(phen), magnesium hydroxide
dro	= dro(ne), hydrocortisone
chlor	= chlor(ine), chlorpheniramine
o	= "o", fexofenadine
thi'	= thi(gh), hydrochlorothiazide
a	= (comm)a, (aur)a, acetaminophen
zide	= z(ipper)(r)ide, hydrochlorothiazide

255. Enala<u>pril</u> (Vasotec)

e	= (n)e(t), etodolac
na'	= na(p), (g)na(t), clonazepam
la	= (co)la, (Koa)la, desvenlafaxine
pril	= pril(l), benazepril

256. Fosinopril (Monopril)

fo	= fo(e), fo(am), fosinopril
si'	= si(t), simethicone
no	= no, (k)no(w), phenobarbital
pril	= pril(l), benazepril

257. Quinapril (Accupril)

qui'	= qui(nce), qui(t), quinapril
na	= (bana)na, fexofenadine
pril	= pril(l), benazepril

258. Lisinopril (Zestril)

li	= li(ght), li(e), lidocaine
si'	= si(t), simethicone
no	= no, (k)no(w), phenobarbital
pril	= pril(l), benazepril

Alternate pronunciation:

li	= li(ly), li(nt), palivizumab
si'	= si(t), simethicone
no	= no, (k)no(w), phenobarbital
pril	= pril(l), benazepril

259. Lisinopril / Hydrochlorothiazide (Zestoretic)

li	= li(ght), li(e), lidocaine
si'	= si(t), simethicone
no	= no, (k)no(w), phenobarbital
pril	= pril(l), benazepril
hy	= hy(brid), hy(phen), magnesium hydroxide
dro	= dro(ne), hydrocortisone
chlor	= chlor(ine), chlorpheniramine
o	= "o", fexofenadine
thi'	= thi(gh), hydrochlorothiazide
a	= (comm)a, (aur)a, acetaminophen
zide	= z(ipper)(r)ide, hydrochlorothiazide

260. Rami<u>pril</u> (Altace)

ra	= ra(ttle), loratadine
mi'	= mi(nt), mi(lk), acetaminophen
pril	= pril(l), benazepril

Angiotensin II receptor blockers (ARBs)

261. Cande<u>sartan</u> (Atacand)

can	= can, can(dy), candesartan
de	= de(sk), de(n), budesonide
sar'	= sar(dine), (pul)sar, candesartan
tan	= tan, tan(trum), candesartan
hy	= hy(brid), hy(phen), magnesium hydroxide
dro	= dro(ne), hydrocortisone
chlor	= chlor(ine), chlorpheniramine
o	= "o", fexofenadine
thi'	= thi(gh), hydrochlorothiazide
a	= (comm)a, (aur)a, acetaminophen
zide	= z(ipper)(r)ide, hydrochlorothiazide

262. Irbe<u>sartan</u> (Avapro)

ir	= (f)ir, (s)ir, irbesartan
be	= be(d), rabeprazole
sar'	= sar(dine), (pul)sar, candesartan
tan	= tan, tan(trum), candesartan

263. Irbe<u>sartan</u> / Hydrochloro<u>thiazide</u> (Avalide)

ir	= (f)ir, (s)ir, irbesartan
be	= be(d), rabeprazole
sar'	= sar(dine), (pul)sar, candesartan
tan	= tan, tan(trum), candesartan
hy	= hy(brid), hy(phen), magnesium hydroxide
dro	= dro(ne), hydrocortisone
chlor	= chlor(ine), chlorpheniramine
o	= "o", fexofenadine
thi'	= thi(gh), hydrochlorothiazide

a   = (comm)a, (aur)a, acetaminophen
zide   = z(ipper)(r)ide, hydrochlorothiazide

264. Lo<u>sartan</u> (Cozaar)

lo   = (Co)lo(rado), (s)lo(w), loperamide
sar'   = sar(dine), (pul)sar, candesartan
tan   = tan, tan(trum), candesartan

265. Lo<u>sartan</u> / Hydrochloro<u>thiazide</u> (Hyzaar)

lo   = (Co)lo(rado), (s)lo(w), loperamide
sar'   = sar(dine), (pul)sar, candesartan
tan   = tan, tan(trum), candesartan

hy   = hy(brid), hy(phen), magnesium hydroxide
dro   = dro(ne), hydrocortisone
chlor   = chlor(ine), chlorpheniramine
o   = "o", fexofenadine
thi'   = thi(gh), hydrochlorothiazide
a   = (comm)a, (aur)a, acetaminophen
zide   = z(ipper)(r)ide, hydrochlorothiazide

266. Olme<u>sartan</u> (Benicar)

ol   = ol(d), olmesartan
me   = me(n), me(sh), esomeprazole
sar'   = sar(dine), (pul)sar, candesartan
tan   = tan, tan(trum), candesartan

267. Olme<u>sartan</u> / HCTZ (Benicar HCT)

ol   = ol(d), olmesartan
me   = me(n), me(sh), esomeprazole
sar'   = sar(dine), (pul)sar, candesartan
tan   = tan, tan(trum), candesartan

hy   = hy(brid), hy(phen), magnesium hydroxide
dro   = dro(ne), hydrocortisone
chlor   = chlor(ine), chlorpheniramine
o   = "o", fexofenadine
thi'   = thi(gh), hydrochlorothiazide

a   = (comm)a, (aur)a, acetaminophen
zide   = z(ipper)(r)ide, hydrochlorothiazide

268. Telmisartan / HCTZ (Micardis HCT)

tel   = tel(ephone), telmisartan
mi   = mi(nt), mi(lk), acetaminophen
sar'   = sar(dine), (pul)sar, candesartan
tan   = tan, tan(trum), candesartan

hy   = hy(brid), hy(phen), magnesium hydroxide
dro   = dro(ne), hydrocortisone
chlor   = chlor(ine), chlorpheniramine
o   = "o", fexofenadine
thi'   = thi(gh), hydrochlorothiazide
a   = (comm)a, (aur)a, acetaminophen
zide   = z(ipper)(r)ide, hydrochlorothiazide

269. Valsartan (Diovan)

val   = val(ley), val(ve), divalproex
sar'   = sar(dine), (pul)sar, candesartan
tan   = tan, tan(trum), candesartan

270. Valsartan / HCTZ (Diovan HCT)

val   = val(ley), val(ve), divalproex
sar'   = sar(dine), (pul)sar, candesartan
tan   = tan, tan(trum), candesartan

hy   = hy(brid), hy(phen), magnesium hydroxide
dro   = dro(ne), hydrocortisone
chlor   = chlor(ine), chlorpheniramine
o   = "o", fexofenadine
thi'   = thi(gh), hydrochlorothiazide
a   = (comm)a, (aur)a, acetaminophen
zide   = z(ipper)(r)ide, hydrochlorothiazide

V. CALCIUM CHANNEL BLOCKERS (CCBs)

Non-dihydropyridines

271. Diltiazem (Cardizem)

dil	= dil(l), dil(uent), diltiazem
ti'	= ti(e), ti(de), tizanidine
a	= (comm)a, (aur)a, acetaminophen
zem	= (ec)zem(a), diltiazem

272. Verapamil (Calan)

ve	= ve(rb), (e)ve(r), verapamil
ra'	= ra(ttle), loratadine
pa	= (pa)pa, levodopa
mil	= mil(k), verapamil

Dihydropyridines

273. Amlodipine (Norvasc)

am	= (l)am(p), triamcinolone
lo'	= (Co)lo(rado), (s)lo(w), loperamide
di	= di(nner), di(n), cefdinir
pine	= (shi)p(mar)ine, atropine

274. Amlodipine / Atorvastatin (Caduet)

am	= (l)am(p), triamcinolone
lo'	= (Co)lo(rado), (s)lo(w), loperamide
di	= di(nner), di(n), cefdinir
pine	= (shi)p(mar)ine, atropine
a	= (comm)a, (aur)a, acetaminophen
tor'	= tor(so), tor(t), atorvastatin
va	= (lar)va, (sali)va, varenicline
sta	= sta(ff), nystatin
tin	= tin, tin(der), nystatin

275. Amlodipine / Benazepril (Lotrel)

am	= (l)am(p), triamcinolone

lo'   = (Co)lo(rado), (s)lo(w), loperamide
di   = di(nner), di(n), cefdinir
pine   = (shi)p(mar)ine, atropine

be   = be(d), rabeprazole
na'   = na(p), (g)na(t), clonazepam
ze   = ze(d), ze(phyr), azelastine
pril   = pril(l), benazepril

276. Amlo<u>dipine</u> / Val<u>sartan</u> (Exforge)

am   = (l)am(p), triamcinolone
lo'   = (Co)lo(rado), (s)lo(w), loperamide
di   = di(nner), di(n), cefdinir
pine = (shi)p(mar)ine, atropine

val   = val(ley), val(ve), divalproex
sar' = sar(dine), (pul)sar, candesartan
tan   = tan, tan(trum), candesartan

277. Felo<u>dipine</u> (Plendil)

fe   = fe(n), fexofenadine
lo'   = (Co)lo(rado), (s)lo(w), loperamide
di   = di(nner), di(n), cefdinir
pine = (shi)p(mar)ine, atropine

278. Nife<u>dipine</u> (Procardia)

ni   = ni(ght), brompheniramine
fe'   = fe(n), fexofenadine
di   = di(nner), di(n), cefdinir
pine = (shi)p(mar)ine, atropine

VI. VASODILATORS

279. Hy<u>dra</u>lazine (Apresoline)

hy   = hy(brid), hy(phen), magnesium hydroxide
dra' = dra(gon), hydralazine
la   = (co)la, (Koa)la, desvenlafaxine

| zine | = (maga)zine, cetirizine |

280. Isosorbide mononitrate (Imdur)

i'	= "i", ibuprofen
so	= so(fa), so, docusate sodium
sor'	= sor(bet), sor(e), isosorbide mononitrate
bide	= bide, (a)bide, isosorbide mononitrate
mo'	= mo(nolith), isosorbide mononitrate
no	= no, (k)no(w), phenobarbital
ni'	= ni(ght), brompheniramine
trate	= (concen)trate, (magis)trate, metoprolol tartrate

281. Nitroglycerin (Nitrostat)

ni	= ni(ght), brompheniramine
tro	= (me)tro, (re)tro, dextromethorphan
gly'	= gly(ph), nitroglycerin
ce	= ce(nt), ce(nter), cetirizine
rin	= rin(se), permethrin

VII. ANTI-ANGINAL

282. Ranolazine (Ranexa)

ra	= (au)ra, (cob)ra, (Op)ra(h), loperamide
no'	= no, (k)no(w), phenobarbital
la	= (co)la, (Koa)la, desvenlafaxine
zine	= (maga)zine, cetirizine

VIII. ANTIHYPERLIPIDEMICS

HMG-CoA reductase inhibitors

283. Atorvastatin (Lipitor)

a	= (comm)a, (aur)a, acetaminophen
tor'	= tor(so), tor(t), atorvastatin
va	= (lar)va, (sali)va, varenicline
sta	= sta(ff), nystatin

tin   = tin, tin(der), nystatin

284. Lovastatin (Mevacor)

lo'	= (Co)lo(rado), (s)lo(w), loperamide
va	= (lar)va, (sali)va, varenicline
sta	= sta(ff), nystatin
tin	= tin, tin(der), nystatin

285. Pravastatin (Pravachol)

pra'	= pra(nce), (so)pra(no), propranolol
va	= (lar)va, (sali)va, varenicline
sta	= sta(ff), nystatin
tin	= tin, tin(der), nystatin

286. Rosuvastatin (Crestor)

ro	= ro(w), ro(ad), mupirocin
su'	= su(e), Su(san), sumatriptan
va	= (lar)va, (sali)va, varenicline
sta	= sta(ff), nystatin
tin	= tin, tin(der), nystatin

287. Simvastatin (Zocor)

sim'	= sim(ple), sim(mer), simvastatin
va	= (lar)va, (sali)va, varenicline
sta	= sta(ff), nystatin
tin	= tin, tin(der), nystatin

Fibric acid derivatives

288. Fenofibrate (Tricor)

fe	= fe(n), fexofenadine
no	= no, (k)no(w), phenobarbital
fi'	= fi(n), fi(b), fidaxomicin
brate	= (vi)brate, (cali)brate, fenofibrate

Alternate pronunciation:

fe	= fe(ed), fe(e), fenofibrate
no	= no, (k)no(w), phenobarbital

fi'   = fi(ve), fi(ber), fenofibrate
brate   = (vi)brate, (cali)brate, fenofibrate

289. Gemfibrozil (Lopid)

gem   = gem, gemfibrozil
fi'   = fi(n), fi(b), fidaxomicin
bro   = bro(ther), gemfibrozil
zil   = (Bra)zil, zil(ch), donepezil

Alternate pronunciation:

gem   = gem, gemfibrozil
fi'   = fi(ve), fi(ber), fenofibrate
bro   = bro(ther), gemfibrozil
zil   = (Bra)zil, zil(ch), donepezil

Bile acid sequestrant

290. Colesevelam (Welchol)

co   = co(ne), codeine
le   = le(tter), celecoxib
se'   = se(t), se(nt), ondansetron
ve   = ve(teran), ve(st), carvedilol
lam   = lam(b), lam(p), alprazolam

Cholesterol absorption blocker

291. Ezetimibe (Zetia)

e   = (n)e(t), etodolac
ze'   = ze(d), ze(phyr), azelastine
ti   = ti(c), (plas)ti(c), cetirizine
mibe   = m(y)(tr)ibe, ezetimibe

292. Ezetimibe / Simvastatin (Vytorin)

e   = (n)e(t), etodolac
ze'   = ze(d), ze(phyr), azelastine
ti   = ti(c), (plas)ti(c), cetirizine
mibe   = m(y)(tr)ibe, ezetimibe

sim'	= sim(ple), sim(mer), simvastatin
va	= (lar)va, (sali)va, varenicline
sta	= sta(ff), nystatin
tin	= tin, tin(der), nystatin

IX. ANTICOAGULANTS AND ANTIPLATELETS

Anticoagulants

293. Enoxaparin (Lovenox)

e	= (n)e(t), etodolac
nox	= nox(ious), (equi)nox, diphenoxylate / atropine
a	= (comm)a, (aur)a, acetaminophen
pa'	= pa(ir), pa(re), enoxaparin
rin	= rin(se), permethrin

294. Heparin

he'	= he(n), he(lp), heparin
pa	= (pa)pa, levodopa
rin	= rin(se), permethrin

295. Warfarin (Coumadin)

war'	= war, war(fare), warfarin
fa	= (so)fa, sulfamethoxazole/trimethoprim
rin	= rin(se), permethrin

296. Dabigatran (Pradaxa)

da	= (so)da, clindamycin
bi'	= bi(t), (or)bi(t), lubiprostone
ga	= (yo)ga, dabigatran
tran	= tran(sfer), tran(smit), dabigatran

Alternate pronunciation:

da	= (so)da, clindamycin
bi	= bi(t), (or)bi(t), lubiprostone
ga'	= ga(s), ga(p), gatifloxacin
tran	= tran(sfer), tran(smit), dabigatran

297. Rivaroxaban (Xarelto)

ri	= ri(de), ri(val), rivaroxaban
va	= (lar)va, (sali)va, varenicline
rox'	= (xe)rox, p(rox)imal, cefuroxime
a	= (comm)a, (aur)a, acetaminophen
ban	= ban(d), ban(ner), ibandronate

Alternate pronunciation:

ri	= ri(nse), (Flo)ri(da), cetirizine
va	= (lar)va, (sali)va, varenicline
rox'	= (xe)rox, p(rox)imal, cefuroxime
a	= (comm)a, (aur)a, acetaminophen
ban	= ban(d), ban(ner), ibandronate

298. Apixaban (Eliquis)

a	= (comm)a, (aur)a, acetaminophen
pix'	= pix(el), pix(ie), apixaban
a	= (comm)a, (aur)a, acetaminophen
ban	= ban(d), ban(ner), ibandronate

Antiplatelet

299. Aspirin / Dipyridamole (Aggrenox)

as'	= as(p), aspirin
pi	= pi(n), aspirin
rin	= rin(se), permethrin

di'	= di(ce), diphenhydramine
py	= py(xis), dipyridamole
ri	= ri(nse), (Flo)ri(da), cetirizine
da'	= (so)da, clindamycin
mole	= mole, dipyridamole

300. Clopidogrel (Plavix)

clo	= clo(th), clo(g), clonazepam
pi'	= pi(n), aspirin
do	= do(ugh), lidocaine
grel	= (mon)grel, clopidogrel

301. Prasugrel (Effient)

pra'	= pra(nce), (so)pra(no), propranolol
su	= su(e), Su(san), sumatriptan
grel	= (mon)grel, clopidogrel

302. Ticagrelor (Brilinta)

ti	= ti(e), ti(de), tizanidine
ca'	= ca(t), ca(p), ticagrelor
gre	= gre(mlin), gre(nade), ticagrelor
lor	= lor(d), lor(e), ticagrelor

X. CARDIAC GLYCOSIDE AND ANTICHOLINERGIC

Cardiac glycoside

303. Digoxin (Lanoxin)

di	= di(nner), di(n), cefdinir
gox'	= (oxy)g(en)(b)ox, digoxin
in	= in, (t)in, polymyxin B

Anticholinergic

304. Atropine (AtroPen)

a'	= (m)a(t), (b)a(t), diphenoxylate / atropine
tro	= (me)tro, (re)tro, dextromethorphan
pine	= (shi)p(mar)ine, atropine

XI. ANTIDYSRHYTHMIC

305. Amiodarone (Cordarone)

a	= (m)a(t), (b)a(t), diphenoxylate / atropine
mi'	= (se)mi, (a)mi, amiodarone [Note: rhymes with me]
o	= "o", fexofenadine
da	= (so)da, clindamycin
rone	= (d)rone, (p)rone, buspirone

CHAPTER 7 - ENDOCRINE / MISC.

MEDICATIONS

I. OTC INSULIN AND EMERGENCY CONTRACEPTION

306. Regular Insulin (Humulin R)

in'	= in, (t)in, polymyxin B
su	= su(n), su(b), insulin
lin	= lin(t), amoxicillin

307. NPH Insulin (Humulin N)

in'	= in, (t)in, polymyxin B
su	= su(n), su(b), insulin
lin	= lin(t), amoxicillin

308. Levonorgestrel (Plan B One-Step)

le	= le(tter), celecoxib
vo	= vo(te), sevoflurane
nor	= nor, nor(th), buprenorphine / naloxone
ges'	= ges(t), (in)ges(t), levonorgestrel
trel	= trel(lis), levonorgestrel

II. DIABETES AND INSULIN

Biguanides

309. Metformin (Glucophage)

met	= (co)met, met(al), metformin
for'	= for, for(t), formoterol
min	= min(t), (vita)min, metformin

310. Metformin / Glyburide (Glucovance)

met	= (co)met, met(al), metformin

for'	= for, for(t), formoterol
min	= min(t), (vita)min, metformin
gly'	= gly(cemic), polyethylene glycol
bu	= bu(reau), ibuprofen
ride	= ride, (p)ride, glyburide

DPP-4 Inhibitors (Gliptins)

311. Linagliptin (Tradjenta)

li	= li(ly), li(nt), palivizumab
na	= (bana)na, fexofenadine
glip'	= (bi)g(s)lip, linagliptin
tin	= tin, tin(der), nystatin

312. Saxagliptin (Onglyza)

sax	= sax(ophone), sax(ony), saxagliptan
a	= (comm)a, (aur)a, acetaminophen
glip'	= (bi)g(s)lip, linagliptin
tin	= tin, tin(der), nystatin

313. Sitagliptin (Januvia)

si	= si(t), simethicone
ta	= (da)ta, (sona)ta, (quo)ta, acetaminophen
glip'	= (bi)g(s)lip, linagliptin
tin	= tin, tin(der), nystatin

Meglitinides (Glinides)

314. Repaglinide (Prandin)

re	= re(d), re(f), varenicline
pa'	= pa(th), palivizumab
gli	= gli(b), gli(nt), glipizide
nide	= (s)nide, (cya)nide, budesonide

Sulfonylureas – 2nd-generation

315. Glyburide (DiaBeta)

gly'	= gly(cemic), polyethylene glycol
bu	= bu(reau), ibuprofen
ride	= ride, (p)ride, glyburide

316. Glimepiride (Amaryl)

gli	= gli(b), gli(nt), glipizide
me'	= me(n), me(sh), esomeprazole
pi	= pi(n), aspirin
ride	= ride, (p)ride, glyburide

317. Glipizide (Glucotrol)

gli'	= gli(b), gli(nt), glipizide
pi	= pi(n), aspirin
zide	= z(ipper)(r)ide, hydrochlorothiazide

Thiazolidinediones (Glitazones)

318. Pioglitazone (Actos)

pi	= pi(e), pioglitazone
o	= "o", fexofenadine
gli'	= gli(b), gli(nt), glipizide
ta	= (da)ta, (sona)ta, (quo)ta, acetaminophen
zone	= zone, (end)zone, pioglitazone

319. Rosiglitazone (Avandia)

ro	= ro(w), ro(ad), mupirocin
si	= si(t), simethicone
gli'	= gli(b), gli(nt), glipizide
ta	= (da)ta, (sona)ta, (quo)ta, acetaminophen
zone	= zone, (end)zone, pioglitazone

Incretin mimetics

320. Exenatide (Byetta)

ex	= ex(po), ex(it), divalproex [Note: not "egg zit"]

e'   = (n)e(t), etodolac
na   = (bana)na, fexofenadine
tide   = tide, (rip)tide, enfuvirtide (T-20)

321. Liraglutide (Victoza)

li   = li(ly), li(nt), palivizumab
ra'   = (au)ra, (cob)ra, (Op)ra(h), loperamide
glu   = glu(ten), glu(e), glucagon
tide   = tide, (rip)tide, enfuvirtide (T-20)

Hypoglycemia

322. Glucagon (GlucaGen)

glu'   = glu(ten), glu(e), glucagon
ca   = (or)ca, fluticasone
gon   = (hexa)gon, gon(e), glucagon

RX Insulin

323. Insulin aspart (Novolog)

in'   = in, (t)in, polymyxin B
su   = su(n), su(b), insulin
lin   = lin(t), amoxicillin

as'   = as(p), aspirin
part   = part, (a)part(ment), insulin aspart

324. Insulin lispro (Humalog)

in'   = in, (t)in, polymyxin B
su   = su(n), su(b), insulin
lin   = lin(t), amoxicillin

lis'   = lis(t), lis(ten), lisdexamfetamine
pro   = pro(ton), pro(be), pro(active), ibuprofen

325. Insulin detemir (Levemir)

in'   = in, (t)in, polymyxin B
su   = su(n), su(b), insulin

lin = lin(t), amoxicillin

de' = de(sk), de(n), budesonide
te = te(nt), motelukast
mir = mir(ror), detemir

Alternate pronunciation:

de' = de(sk), de(n), budesonide
te = te(nt), motelukast
mir = (ad)mir(al), mir(th), mirtazapine

326. Insulin glargine (Lantus, Toujeo)

in' = in, (t)in, polymyxin B
su = su(n), su(b), insulin
lin = lin(t), amoxicillin

glar' = g(old)lar(k), g(ood)lar(d), glargine
gine = (aspara)gine, lamotrigine

III. THYROID HORMONES

Hypothyroidism

327. Levothyroxine (Synthroid)

le = le(an), levothyroxine
vo = vo(te), sevoflurane
thy = thy(roid), thy(mus), levothyroxine
rox' = (xe)rox, p(rox)imal, cefuroxime
ine = (fam)ine, (eng)ine, levothyroxine

Alternate pronunciation:

le = le(tter), celecoxib
vo = vo(te), sevoflurane
thy = thy(roid), thy(mus), levothyroxine
rox' = (xe)rox, p(rox)imal, cefuroxime
ine = (sal)ine, desvenlafaxine

Hyperthyroidism

328. Propylthiouracil (PTU)

pro	= pro(ton), pro(be), pro(active), ibuprofen
pyl	= (po)p(vin)yl, propylthiouracil
thi	= thi(gh), propylthiouracil
o	= "o", fexofenadine
u'	= "u", cefuroxime
ra	= (au)ra, (cob)ra, (Op)ra(h), loperamide
cil	= (pen)cil, amoxicillin

IV. HORMONES AND CONTRACEPTION

Low testosterone

329. Testosterone (AndroGel)

tes	= tes(t), tes(tudo), testosterone
to'	= to(t), to(p), testosterone
ste	= ste(ps), ste(ms), testosterone
rone	= (d)rone, (p)rone, buspirone

Estrogens and / or progestins

330. Estradiol (Estrace, Estraderm)

es	= (m)es(s), Es(ther), esomeprazole
tra	= (orches)tra, (spec)tra, (ex)tra, sertraline
di'	= di(ce), diphenhydramine
ol	= (aw)ol, (alcoh)ol, albuterol

331. Conjugated estrogens (Premarin)

es'	= (m)es(s), Es(ther), esomeprazole
tro	= (me)tro, (re)tro, dextromethorphan
gens	= (oxy)gens, estrogens

332. Conjugated estrogens / Medroxyprogesterone (Prempro, Premphase)

es'	= (m)es(s), Es(ther), esomeprazole

tro   = (me)tro, (re)tro, dextromethorphan
gens   = (oxy)gens, estrogens

me   = me(n), me(sh), esomeprazole
drox   = dr(ink)(b)ox, dr(op)(b)ox, magnesium hydroxide
y   = (countr)y, oxycodone
pro   = pro(ton), pro(be), pro(active), ibuprofen
ges'   = ges(t), (in)ges(t), levonorgestrel
te   = te(nt), motelukast
rone   = (d)rone, (p)rone, buspirone

333. Progesterone (Prometrium)

pro   = pro(ton), pro(be), pro(active), ibuprofen
ges'   = ges(t), (in)ges(t), levonorgestrel
te   = te(nt), motelukast
rone   = (d)rone, (p)rone, buspirone

334. Medroxyprogesterone (Provera)

me   = me(n), me(sh), esomeprazole
drox   = dr(ink)(b)ox, dr(op)(b)ox, magnesium hydroxide
y   = (countr)y, oxycodone
pro   = pro(ton), pro(be), pro(active), ibuprofen
ges'   = ges(t), (in)ges(t), levonorgestrel
te   = te(nt), motelukast
rone   = (d)rone, (p)rone, buspirone

Combined oral contraceptive pill (COCP)

335. Ethinyl estradiol / norethindrone / Fe
(Loestrin 24 Fe)

eth'   = (m)eth(od), polyethylene glycol
i   = i(t), simethicone
nyl   = (vi)nyl, fentanyl

es   = (m)es(s), Es(ther), esomeprazole
tra   = (orches)tra, (spec)tra, (ex)tra, sertraline
di'   = di(ce), diphenhydramine
ol   = (aw)ol, (alcoh)ol, albuterol

nor   = nor, nor(th), buprenorphine / naloxone
eth'   = (m)eth(od), polyethylene glycol
in   = in, (t)in, polymyxin B
drone   = drone, norethindrone

336. Ethinyl estradiol / norgestimate (Tri-Sprintec)

eth'   = (m)eth(od), polyethylene glycol
i   = i(t), simethicone
nyl   = (vi)nyl, fentanyl

es   = (m)es(s), Es(ther), esomeprazole
tra   = (orches)tra, (spec)tra, (ex)tra, sertraline
di'   = di(ce), diphenhydramine
ol   = (aw)ol, (alcoh)ol, albuterol

nor   = nor, nor(th), buprenorphine / naloxone
ges'   = ges(t), (in)ges(t), levonorgestrel
ti   = ti(c), (plas)ti(c), cetirizine
mate   = mate, (room)mate, topiramate

Patch

337. Ethinyl estradiol / norelgestromin (OrthoEvra)

eth'   = (m)eth(od), polyethylene glycol
i   = i(t), simethicone
nyl   = (vi)nyl, fentanyl

es   = (m)es(s), Es(ther), esomeprazole
tra   = (orches)tra, (spec)tra, (ex)tra, sertraline
di'   = di(ce), diphenhydramine
ol   = (aw)ol, (alcoh)ol, albuterol

nor   = nor, nor(th), buprenorphine / naloxone
el   = el(bow), el(ect), eletriptan
ges'   = ges(t), (in)ges(t), levonorgestrel
tro   = (me)tro, (re)tro, dextromethorphan
min   = min(t), (vita)min, metformin

Ring

338. Ethinyl estradiol / etonogestrel (NuvaRing)

eth'	= (m)eth(od), polyethylene glycol
i	= i(t), simethicone
nyl	= (vi)nyl, fentanyl

es	= (m)es(s), Es(ther), esomeprazole
tra	= (orches)tra, (spec)tra, (ex)tra, sertraline
di'	= di(ce), diphenhydramine
ol	= (aw)ol, (alcoh)ol, albuterol

e	= (n)e(t), etodolac
to	= (pis)to(n), (a)to(m), ketoconazole
no	= no, (k)no(w), phenobarbital
ges'	= ges(t), (in)ges(t), levonorgestrel
trel	= trel(lis), levonorgestrel

V. OVERACTIVE BLADDER, URINARY RETENTION, ERECTILE DYSFUNCTION (ED), BENIGN PROSTATIC HYPERPLASIA (BPH)

Overactive bladder

339. Oxybutynin (Ditropan)

ox	= ox, (b)ox, sulfamethoxazole/trimethoprim
y	= (countr)y, oxycodone
bu'	= bu(reau), ibuprofen
ty	= ty(pical), ty(mpanic), amitriptyline
nin	= (mela)nin, oxybutynin

340. Darifenacin (Enablex)

da	= da(rt), (ra)da(r), darifenacin
ri	= ri(nse), (Flo)ri(da), cetirizine
fe'	= fe(n), fexofenadine
na	= (bana)na, fexofenadine
cin	= cin(der), (s)cin(tillate), bacitracin

341. Solifenacin (VESIcare)

so	= so(fa), so, docusate sodium
li	= li(ly), li(nt), palivizumab
fe'	= fe(n), fexofenadine
na	= (bana)na, fexofenadine
cin	= cin(der), (s)cin(tillate), bacitracin

342. Tolterodine (Detrol)

tol	= tol(erate), (derma)tol(ogy), ethambutol
te'	= te(nt), motelukast
ro	= ro(w), ro(ad), mupirocin
dine	= (bir)di(e)(tu)ne, fexofenadine

Urinary retention

343. Bethanechol (Urecholine)

be	= be(d), rabeprazole
tha'	= tha(tch), tha(t), chlorthalidone
ne	= ne(t), guaifenesin
chol	= (s)chol(ar), chol(era), bethanechol

Erectile dysfunction - PDE-5 inhibitors

344. Sildenafil (Viagra)

sil	= sil(ver), sil(l), sildenafil
de'	= de(sk), de(n), budesonide
na	= (bana)na, fexofenadine
fil	= fil(m), fil(l), sildenafil

345. Vardenafil (Levitra)

var	= var(sity), var(nish), vardenafil
de'	= de(sk), de(n), budesonide
na	= (bana)na, fexofenadine
fil	= fil(m), fil(l), sildenafil

346. Tadalafil (Cialis)

ta	= (da)ta, (sona)ta, (quo)ta, acetaminophen
da'	= da(nce), tadalafil

la   = (co)la, (Koa)la, desvenlafaxine
fil   = fil(m), fil(l), sildenafil

BPH – Alpha-blocker

347. Alfuzosin (Uroxatral)
al   = (s)al(iva), Al(exander), albuterol
fu'   = fu(mes), fu(ry), enfuvirtide (T-20)
zo   = zo(mbie), benzonatate
sin   = (ba)sin, sin, guaifenesin

348. Tamsulosin (Flomax)
tam   = tam(ale), levetiracetam
su'   = su(e), Su(san), sumatriptan
lo   = (Co)lo(rado), (s)lo(w), loperamide
sin   = (ba)sin, sin, guaifenesin

BPH – 5-alpha-reducase inhibitor

349. Dutasteride (Avodart)
du   = du(et), du(ne), duloxetine
ta'   = ta(b), ta(x), Oxymetazoline
ste   = ste(ps), ste(ms), testosterone
ride   = ride, (p)ride, glyburide

350. Finasteride (Proscar, Propecia)
fi   = fi(n), fi(b), fidaxomicin
na'   = na(p), (g)na(t), clonazepam
ste   = ste(ps), ste(ms), testosterone
ride   = ride, (p)ride, glyburide

CHAPTER 8 - PEDIATRIC ONCOLOGY MEDICATIONS

The challenges parents in the NICU have in some ways parallel those of pediatric cancer patients – long stays, uncertainty, and many medications. As oncology is a pathophysiologic state we haven't covered, here we'll expect to add some new sounds while still reusing many that we've heard before.

Leukemia – cancers of the bone marrow and blood – are the most common in children. Leukemia accounts for about 30% of pediatric cancers in the U.S. Other forms of cancer in children include bone and brain cancers, lymphoma (Hodgkin and non-Hodgkin), neuroblastoma, retinoblastoma, rhabdomyosarcoma, and Wilms' tumor.

I want to talk a little bit about what made me add this section. At the Examined Life Conference in Iowa City, Iowa, I met many doctors, nurses, and pharmacists who expressed their own and their patients' stories. I wanted to make sure I addressed the disease that so deeply impacted this group.

Please note that descriptions are here for context. I'm not making any medical recommendations. **I put a new medication syllable in bold on first use.**

PEDIATRIC ONCOLOGY MEDICATIONS

351. Asparaginase (Erwinaze): Miscellaneous injectable chemotherapy drug, an enzyme that depletes asparagine.

a	= (m)a(t), (b)a(t), diphenoxylate / atropine
spa	= **spa(re), asparaginase**
ra¹	= (au)ra, (cob)ra, (Op)ra(h), loperamide
gi	= (ma)gi(c), (lo)gi(c), selegiline

nase   = (oxyge)nase, asparaginase

352. Bleomycin (Blenoxane): Antitumor antibiotic, nonanthrocycline.

ble   **= ble(ep), ble(ak), bleomycin**
o   = "o", fexofenadine
my'   = my, my(osin), neomycin sulfate
cin   = cin(der), (s)cin(tillate), bacitracin

353. Busulfan (Myleran): Alkylating agent, other.

bu   = bu(reau), ibuprofen
sul'   = sul(len), neomycin sulfate
fan   **= (in)fan(t), busulfan**

354. Cisplatin (Platinol): Platinum compound.

cis'   **= (public)cis(t), cisplatin**
pla   **= pla(tinum), cisplatin**
tin   = tin, tin(der), nystatin

355. Cyclophosphamide (Cytoxan, Neosar): Alkylating agent, nitrogen mustard.

cy   = cy(cle), cy(clone), cyclobenzaprine
clo   = clo(ver), clo(the), diclofenac
phos'   **= phos(phor), cyclophosphamide**
pha   = (al)pha, cephalexin
mide   = (gu)m(sl)ide, loperamide

356. Cyclosporine (Sandimmune Injection): Immunosuppressant drug.

cy   = cy(cle), cy(clone), cyclobenzaprine
clo   = clo(ver), clo(the), diclofenac
spo'   **= spo(rt), cyclosporine**
rine   = (doct)rine, (u)rine, phenylephrine

357. Cytarabine (Cytosar-U): Antimetabolite, pyrimidine analog.

cy   = cy(st), bismuth subsalicylate
ta'   = ta(b), ta(x), Oxymetazoline
ra   = (au)ra, (cob)ra, (Op)ra(h), loperamide
bine   = (zom)bi(e)(li)ne, (Yohim)bine, emtricitabine

358. Daunorubicin (Cerubidine): Antitumor antibiotic, anthracycline.

dau	= dau(nt), daunorubicin
no	= (can)no(n), (ca)no(n), acetaminophen
ru'	= ru(by), ru(in), darunavir
bi	= bi(t), (or)bi(t), lubiprostone
cin	= cin(der), (s)cin(tillate), bacitracin

359. Dactinomycin (Cosmegan): Antitumor antibiotic, anthracycline.

dac	= (di)dac(tic), dac(tyl), dactinomycin
ti	= ti(c), (plas)ti(c), cetirizine
no	= (can)no(n), (ca)no(n), acetaminophen
my'	= my, my(osin), neomycin sulfate
cin	= cin(der), (s)cin(tillate), bacitracin

360. Doxorubicin (Adriamycin, Rubex): Antitumor antibiotic, anthracycline.

dox	= (para)dox, (ortho)dox, doxycycline
o	= "o", fexofenadine
ru'	= ru(by), ru(in), darunavir
bi	= bi(t), (or)bi(t), lubiprostone
cin	= cin(der), (s)cin(tillate), bacitracin

361. Dexamethasone (Decadron): An an anti-inflammatory or for nausea.

dex	= dex(terity), Dex(ter), dextromethorphan
a	= (comm)a, (aur)a, acetaminophen
meth'	= meth(od), simethicone
a	= (comm)a, (aur)a, acetaminophen
sone	= so(fa), (du)ne, fluticasone

362. Dexrazoxane (Zinecard, Totect): Chemoprotectant drug sometimes given with doxorubicin to stop pediatric cardiomyopathy.

dex'	= dex(terity), Dex(ter), dextromethorphan
ra	= (au)ra, (cob)ra, (Op)ra(h), loperamide
zox'	= (qui)z(b)ox, nitazoxanide
ane	= (c)ane, sevoflurane

363. Etoposide (Toposar): Topoisomerase inhibitor.

e	= (n)e(t), etodolac
to	= to(t), to(p), testosterone
po'	= po(t), po(lygon), polyethylene glycol
side	= **side, (out)side, etoposide**

364. Granisetron (Kytril): Strong anti-nausea medicine, technically a $5\text{-}HT_3$ receptor antagonist, oral dosing, for chemotherapy induced nausea and vomiting, CINV.

gra	= (ag)gra(vate), raltegravir
ni'	= (k)ni(t), prednisolone
se	= se(t), se(nt), ondansetron
tron	= (elec)tron, ondansetron

365. Imatinib (Gleevec): BCR-ABL Tyrosine kinase inhibitor.

i	= i(t), simethicone
ma'	= ma(p), ma(t), temazepam
ti	= ti(c), (plas)ti(c), cetirizine
nib	= **nib(ble), imatinib**

366. Irinotecan (Camptosar): Topoisomerase inhibitor.

i	= i(t), simethicone
ri	= ri(nse), (Flo)ri(da), cetirizine
no	= no(d), no(t), tenofovir
te'	= te(a), te(e), ramelteon
can	= can, can(dy), candesartan

Alternate:

i	= "i", ibuprofen
ri	= ri(nse), (Flo)ri(da), cetirizine
no	= no(d), no(t), tenofovir
te'	= te(a), te(e), ramelteon
can	= can, can(dy), candesartan

367. Mercaptopurine (Purinethol, Purixan): Antimetabolite, purine analog.

mer	= (ga)mer, mercaptopurine
cap	= cap, cap(itol), mercaptopurine
to	= to(e), to(w), pantoprazole
pur'	= pur(e), pur(ity), allopurinol
ine	= (sal)ine, desvenlafaxine

368. Methotrexate (Abitrexate): Antimetabolite, folic acid analog.

meth	= meth(od), simethicone
o	= "o", fexofenadine
trex'	= tre(mor), "x", methotrexate
ate	= (g)ate, (l)ate, diphenoxylate / atropine

369. Rituximab (Rituxan): CD20-directed antibody.

ri	= ri(nse), (Flo)ri(da), cetirizine
tux'	= tux(edo), rituximab
i	= i(t), simethicone
mab	= m(at)(c)ab, palivizumab

370. Vincristine (VCR, Oncovin, Vincasar): Mitotic inhibitor, vinca alkaloid.

vin	= vin(tage), vincristine
cris'	= cris(py), vincristine
tine	= (sal)tine, (rou)tine, fluoxetine

CHAPTER 9 - 30 RECENT ADDITIONS

As we add new medications, how many new syllables do we need to add? Have we learned enough? I think it's reasonable to predict that, as new drugs come out, we'll already know most of the syllables or more. This makes our job much easier from here on out. **I put a new medication syllable in bold on first use.**

RESPIRATORY

371. Aclidinium bromide (Tudorza Pressair): Anticholinergic COPD medication.

a	= (m)a(t), (b)a(t), diphenoxylate / atropine
cli	= cli(ff), meclizine
di'	= di(nner), di(n), cefdinir
ni	**= (mi)ni, aclidinium bromide**
um	= (g)um, calcium carbonate

372. Umeclidinium bromide / vilanterol (Anoro Ellipta):
Combination long-acting anticholinergic and ultra-long-acting Beta$_2$ agonist for COPD.

u	= "u", cefuroxime
me	= me(n), me(sh), esomeprazole
cli	= cli(ff), meclizine
di'	= di(nner), di(n), cefdinir
ni	= (mi)ni, aclidinium bromide
um	= (g)um, calcium carbonate

373. Fluticasone furoate / vilanterol (Breo Ellipta): Inhaled corticosteroid and ultra-long-acting Beta$_2$ agonist for COPD.

flu	= flu(id), fluticasone
ti'	= ti(c), (plas)ti(c), cetirizine
ca	= (or)ca, fluticasone
sone	= so(fa),(du)ne, fluticasone

fu'	= fu(mes), fu(ry), enfuvirtide (T-20)
ro	= ro(w), ro(ad), mupirocin
ate	= (g)ate, (l)ate, diphenoxylate / atropine
vi	= vi(sion), vi(m), palivizumab
lan'	= lan(ce), lan(d), dexlansoprazole
ter	= ter(se), albuterol
ol	= (aw)ol, (alcoh)ol, albuterol

374. Indacaterol (Arcapta Neohaler): Ultra-long-acting beta receptor agonist dosed once daily.

in	= in, (t)in, polymyxin B
da	= (so)da, clindamycin
ca'	= (or)ca, fluticasone
ter	= ter(se), albuterol
ol	= (aw)ol, (alcoh)ol, albuterol

375. Roflumilast (Daliresp): Long acting phosphodiesterase-4 (PDE-4) inhibitor for COPD.

ro	= ro(w), ro(ad), mupirocin
flu'	= flu(id), fluticasone
mi	= mi(nt), mi(lk), acetaminophen
last	**= last, roflumilast**

IMMUNE

376. Boceprevir (Victrelis): Protease inhibitor antiviral for Hepatitis C.

bo	= bo(ne), calcium, carbonate
ce'	= ce(nt), ce(nter), cetirizine
pre	= pre(dator), pre(dict), buprenorphine.
vir	= vir(ulent), oseltamivir [rhymes with veer]

377. Ledipasvir / sofosbuvir (Harvoni): Two drug combination for Hepatitis C.

le	= le(tter), celecoxib
di'	= di(nner), di(n), cefdinir
pas	**= pas(s), pas(t), ledipasvir**
vir	= vir(ulent), oseltamivir [rhymes with veer]

so	= so(fa), so, docusate sodium
fos'	**= fossil, sofosbuvir**
bu	= bu(reau), ibuprofen
vir	= vir(ulent), oseltamivir [rhymes with veer]

378. Ombit<u>as</u>vir / parit<u>a</u>previr / rit<u>o</u>navir (Viekira Pak): An antiviral triple combination for Hepatitis C.

om	**= (m)om, ombitasvir**
bi'	= bi(t), (or)bi(t), lubiprostone
tas	= tas(k), potassium
vir	= vir(ulent), oseltamivir [rhymes with veer]

pa	= (pa)pa, levodopa
ri	= ri(nse), (Flo)ri(da), cetirizine
ta'	= (da)ta, (sona)ta, (quo)ta, acetaminophen
pre	= pre(dator), pre(dict), buprenorphine / naloxone
vir	= vir(ulent), oseltamivir [rhymes with veer]

ri	= ri(nse), (Flo)ri(da), cetirizine
to'	= to(t), to(p), testosterone
na	= (bana)na, fexofenadine
vir	= vir(ulent), oseltamivir [rhymes with veer]

379. Sime<u>previr</u> (Olysio): Treatment for Hepatitis C.

si	= si(t), simethicone
me'	= me(n), me(sh), esomeprazole
pre	= pre(dator), pre(dict), buprenorphine / naloxone
vir	= vir(ulent), oseltamivir [rhymes with veer]

380. Sofos<u>buvir</u> (Sovaldi): Treatment for Hepatits C.

| so | = so(fa), so, docusate sodium |
| fos' | = fossil, sofosbuvir |

bu	= bu(reau), ibuprofen
vir	= vir(ulent), oseltamivir [rhymes with veer]

381. Tela<u>previr</u> (Incivek and Incivo): Protease inhibitor antiviral for Hepatitis C.

te	= te(nt), motelukast
la'	= la(b), la(st), azelastine
pre	= pre(dator), pre(dict), buprenorphine / naloxone
vir	= vir(ulent), oseltamivir [rhymes with veer]

NEURO

382. Brex<u>pip</u>razole (Rexulti): Atypical antipsychotic drug that is a D_2 partial agonist, a serotonin-dopamine activity modulator (SDAM) for schizophrenia.

brex	= **Brexit, brexipiprazole**
pi'	= pi(n), aspirin
pra	= (su)pra, (O)pra(h), dexlansoprazole
zole	= (fe)z(p)ole, sulfamethoxazole/trimethoprim

383. Cari<u>praz</u>ine (Vraylar): Atypical antipsychotic that works as a partial agonist on D2 and D3 receptors, with D3 selectivity for schizophrenia.

ca	= (or)ca, fluticasone
ri'	= ri(nse), (Flo)ri(da), cetirizine
pra	= (su)pra, (O)pra(h), dexlansoprazole
zine	= (maga)zine, cetirizine

384. Suv<u>orex</u>ant (Belsomra): Sedative / hypnotic that works as a selective, dual orexin receptor antagonist.

su'	= su(e), Su(san), sumatriptan
vo	= vo(te), sevoflurane
rex'	= **(ano)rex(ic), suvorexant**
ant	= **(gi)ant, suvorexant**

CARDIOVASCULAR

385. Aliskiren (Tekturna): Antihypertensive, direct renin inhibitor.

a	= (m)a(t), (b)a(t), diphenoxylate / atropine
lis	= lis(t), lis(ten), lisdexamfetamine
ki'	= **ki(d), aliskiren**
ren	= (si)ren, ren(t), aliskiren

386. Alirocumab (Praluent): Proprotein convertase subtilisin / kexin type 9 (PCSK9) inhibitor for cholesterol.

a	= (m)a(t), (b)a(t), diphenoxylate / atropine
li	= li(ly), li(nt), palivizumab
ro'	= ro(w), ro(ad), mupirocin
cu	= cu(be), cu(re), cu(te), docusate sodium
mab	= m(at)(c)ab, palivizumab

387. Evolocumab (Repatha): Proprotein convertase subtilisin / kexin type 9 (PCSK9) inhibitor for cholesterol.

e	= (n)e(t), etodolac
vo	= vo(te), sevoflurane
lo'	= (Co)lo(rado), (s)lo(w), loperamide
cu	= cu(be), cu(re), cu(te), docusate sodium
mab	= m(at)(c)ab, palivizumab

388. Idarucizumab (Praxbind): Reverses anticoagulant effects of dabigatran (Pradaxa).

i	= "i", ibuprofen
da	= da(rt), (ra)da(r), darifenacin
ru	= ru(by), ru(in), darunavir
ci'	= (s)ci(ntillate), triamcinolone
zu	= Zu(lu), (shiat)zu, palivizumab
mab	= m(at)(c)ab, palivizumab

389. Sacubi<u>tril</u> / val<u>sar</u>tan (Entresto):
Neprilysin inhibitor / Angiotensin II receptor blocker (ARB)

sa	= (vi)sa, sa(liva), bismuth subsalicylate
cu'	= cu(be), cu(re), cu(te), docusate sodium
bi	= bi(t), (or)bi(t), lubiprostone
tril	**= (nos)tril, tril(l), sacubitril / valsartan**
val	= val(ley), val(ve), divalproex
sar'	= sar(dine), (pul)sar, candesartan
tan	= tan, tan(trum), candesartan

ENDOCRINE – WEIGHT LOSS

390. Lor<u>ca</u>serin (Belviq): Serotonergic weight loss drug.

lor	= lor(d), lor(e), ticagrelor
ca'	= (or)ca, fluticasone
se	= se(t), se(nt), ondansetron
rin	= rin(se), permethrin

391. Phentermine and topiramate (Qsymia): Combination weight loss medication.

phen'	= (hy)phen, acetaminophen
ter	= ter(se), albuterol
mine	= (hista)mine, brompheniramine
to	= to(e), to(w), pantoprazole
pi'	= pi(ece), ipratropium
ra	= (au)ra, (cob)ra, (Op)ra(h), loperamide
mate	= mate, (room)mate, topiramate

392. Pyridoxine/doxylamine (Diclegis): Combination of vitamin B6 and doxylamine succinate for morning sickness.

py	= py(xis), dipyridamole
ri	= ri(nse), (Flo)ri(da), cetirizine
dox'	= (para)dox, (ortho)dox, doxycycline

ine   = (sal)ine, desvenlafaxine

dox   = (para)dox, (ortho)dox, doxycycline
yl'   = (s)yl(lable), polyethylene glycol
a   = (comm)a, (aur)a, acetaminophen
mine   = (hista)mine, brompheniramine

ENDOCRINE – DIABETES

393. Inhalable insulin (Afrezza): Inhalable insulin.

in'   = in, (t)in, polymyxin B
su   = su(n), su(b), insulin
lin   = lin(t), amoxicillin

394. Insulin glargine (Toujeo): Long-acting once daily basal insulin analog.

in'   = in, (t)in, polymyxin B
su   = su(n), su(b), insulin
lin   = lin(t), amoxicillin

glar'   = g(old)lar(k), g(ood)lar(d), glargine
gine   = (aspara)gine, lamotrigine

395. Insulin degludec (Degludec): Ultralong-acting basal insulin analog.

in'   = in, (t)in, polymyxin B
su   = su(n), su(b), insulin
lin   = lin(t), amoxicillin

de   = de(sk), de(n), budesonide
glu'   = glu(ten), glu(e), glucagon
dec   **= dec(k), insulin degludec**

396. Canagliflozin (Invokana): Gliflozin class antidiabetic.

ca   = ca(t), ca(p), ticagrelor
na   = (bana)na, fexofenadine

gli   = gli(b), gli(nt), glipizide
flo'   = **flo(w), flo(at), canagliflozin**
zin   = zin(fandel), canagliflozin

397. Dapagliflozin (Farxiga): Gliflozin class antidiabetic.

da   = da(nce), tadalafil
pa   = (pa)pa, levodopa
gli   = gli(b), gli(nt), glipizide
flo'   = flo(w), flo(at), canagliflozin
zin   = zin(fandel), canagliflozin

398. Empagliflozin (Jardiance): Gliflozin class antidiabetic.

em   = (g)em, em(ber), emtricitabine
pa   = (pa)pa, levodopa
gli   = gli(b), gli(nt), glipizide
flo'   = flo(w), flo(at), canagliflozin
zin   = zin(fandel), canagliflozin

399. Albiglutide (Tanzeum): Glucagon-like peptide agonist (GLP-1 agonist)

al   = (s)al(iva), Al(exander), albuterol
bi   = bi(t), (or)bi(t), lubiprostone
glu'   = glu(ten), glu(e), glucagon
tide   = tide, (rip)tide, enfuvirtide (T-20)

400. Exenatide (Bydureon): Glucagon-like peptide agonist (GLP-1 agonist)

ex   = ex(po), ex(it), divalproex
e'   = (n)e(t), etodolac
na   = (bana)na, fexofenadine
tide   = tide, (rip)tide, enfuvirtide (T-20)

Chapter 10 - The Generic to English translation system (GETS) syllable list

After months of work, late night writing sessions, and many dead ends, this is the product. An easy way for you to look at each new medication you see and translate it into common English words. I feel like I should say something profound, but I've just worked so hard for so long that I simply hope you'll take the time to contact me by writing a review, or contact me personally and let me know if this made you more confident as a patient, a parent, or a provider.

a	= (comm)a, (aur)a, acetaminophen
a	= (m)a(t), (b)a(t), diphenoxylate / atropine
a	= "a", diazepam
a	= a(ir), aripiprazole
al	= (s)al(iva), Al(exander), albuterol
am	= (l)am(p), triamcinolone
an	= an, an(t), nitrofurantoin
ane	= (c)ane, sevoflurane
ant	= (gi)ant, suvorexant
as	= as(p), aspirin
ate	= (g)ate, (l)ate, diphenoxylate / atropine
ax	= ax, (t)ax(i), ceftriaxone
B	= "B"
ba	= (tu)ba, methocarbamol
ba	= ba(t), bacitracin
ba	= ba(y), ba(it), oxcarbazepine
ban	= ban(d), ban(ner), ibandronate
bar	= bar, bar(bell), phenobarbital
be	= be(d), rabeprazole
be	= be(ta), (o)be(y), betamethasone
ben	= ben(d), benzocaine

benz	= (Mercedes)Benz, benz(ene), benztropine
bi	= bi(t), (or)bi(t), lubiprostone
bide	= bide, (a)bide, isosorbide mononitrate
bine	= (zom)bi(e)(li)ne, (Yohim)bine, emtricitabine
bis	= bis(quit), Bis(marck), bismuth subsalicylate
ble	= ble(ep), ble(ak), bleomycin
bo	= bo(ne), calcium, carbonate
box	= box, isocarboxazid
brate	= (vi)brate, (cali)brate, fenofibrate
brex	= Brexit, brexpiprazole
bro	= bro(ther), gemfibrozil
brom	= brom(ine), brompheniramine
bu	= bu(reau), ibuprofen
bux	= bux(om), febuxostat
ca	= (or)ca, fluticasone
ca	= ca(t), ca(p), ticagrelor
caf	= caf(e), caffeine
caine	= c(at)(migr)aine, benzocaine
cal	= cal(orie), (lo)cal(e), calcium carbonate
cam	= cam(p), meloxicam
can	= can, can(dy), candesartan
cap	= cap, cap(itol), mercaptopurine
car	= car, car(d), calcium carbonate
ce	= ce(iling), acetaminophen
ce	= ce(nt), ce(nter), cetirizine
cef	= c(l)ef, c(l)ef(t), cefdinir
cept	= (ac)cept, (abata)cept
chi	= chi(n), chi(p), colchicine
chlor	= chlor(ine), chlorpheniramine
chol	= (s)chol(ar), chol(era), bethanechol
ci	= (pronun)ci(ation), calcium carbonate
ci	= (s)ci(ntillate), triamcinolone
ci	= ci(der), (s)ci(ence), emtricitabine
cil	= (pen)cil, amoxicillin
cin	= cin(der), (s)cin(tillate), bacitracin
cine	= (leu)cine, influenza vaccine
cis	= (public)cis(t), cisplatin

cla	= cla(rinet), cla(ret), clarithromycin
cla	= cla(w), clavulanate
cli	= cli(ff), meclizine
clin	= clin(ic), clindamycin
cline	= cl(iff)(sal)ine, varenicline
cline	= cli(ff)(vi)ne, doxycycline
clo	= clo(th), clo(g), clonazepam
clo	= clo(ver), clo(the), diclofenac
clone	= clone, (cy)clone, eszopiclone
co	= co(ne), codeine
col	= col(lar), Col(orado), polyethylene
cone	= cone, (pine)cone, simethicone
cor	= cor(n), cor(k), hydrocortisone
cox	= cox(swain), celecoxib
cris	= cris(py), vincristine
cu	= cu(be), cu(re), cu(te), docusate sodium
cy	= cy(cle), cy(clone), cyclobenzaprine
cy	= cy(st), bismuth subsalicylate
da	= (so)da, clindamycin
da	= da(nce), tadalafil
da	= da(rt), (ra)da(r), darifenacin
da	= da(y), da(te), midazolam
dac	= (di)dac(tic), dac(tyl), dactinomycin
dan	= dan(ce), Dan, ondansetron
date	= date, (up)date, dexmethylphenidate
dau	= dau(nt), daunorubicin
dax	= (a)dax(ial), (ad)dax, fidaxomicin
de	= de(sk), de(n), budesonide
dec	= dec(k), insulin degludec
deine	= de(er), (mach)ine, codeine
dem	= dem(ocracy), dem(and), zolpidem
des	= des(ks), des(tination), desvenlafaxine
dex	= dex(terity), Dex(ter), dextromethorphan
di	= (bir)di(e), (per)di(em), docusate sodium
di	= di(ce), diphenhydramine
di	= di(nner), di(n), cefdinir
dil	= dil(l), dil(uent), diltiazem

dine	= (bir)di(e)(tu)ne, fexofenadine
do	= do(t), do(dge), docusate sodium
do	= do(ugh), lidocaine
do	= (i)do(l), etodolac
dol	= dol(l), tramadol
done	= (con)done, hydrocodone
dox	= (para)dox, (ortho)dox, doxycycline
dra	= (hy)dra, diphenhydramine
dra	= dra(gon), hydralazine
drine	= (alexan)drine, pseudoephedrine
dro	= dro(ne), hydrocortisone
drone	= drone, norethindrone
drox	= dr(ink)(b)ox, dr(op)(b)ox, magnesium hydroxide
du	= du(et), du(ne), duloxetine
e	= (en)e(my), erythromycin
e	= (n)e(t), etodolac
el	= el(bow), el(ect), eletriptan
em	= (g)em, em(ber), emtricitabine
en	= (p)en, (m)en, influenza vaccine
ene	= (sc)ene, polyethylene glycol
enz	= enz(yme), (fr)enz(y), efavirenz
eph	= (z)eph(yr), phenylephrine
es	= (m)es(s), Es(ther), esomeprazole
eth	= (m)eth(od), polyethylene glycol
ex	= ex(po), ex(it), divalproex
fa	= (so)fa, sulfamethoxazole/trimethoprim
fa	= fa(ct), fa(ctor), efavirenz
fac	= fac(t), fac(tor), surfactant
fam	= fam(ily), fam(ine), rifampin
fan	= (in)fan(t), busulfan
fate	= fate, neomycin sulfate
fax	= fax, (tele)fax, desvenlafaxine
fe	= fe(n), fexofenadine
fe	= fe(ed), fe(e), fenofibrate
feine	= fe(et)(magaz)ine, caffeine
fen	= fen, fen(der), ibuprofen

fene	= fe(et)(du)ne, raloxifene
fex	= (ponti)fex, fexofenadine
fi	= fi(n), fi(b), fidaxomicin
fi	= fi(ve), fi(ber), fenofibrate
fil	= fil(m), fil(l), sildenafil
fine	= fi(eld)(li)ne, butenafine
flix	= fl(our)(m)ix, infliximab
flo	= flo(w), flo(at), canagliflozin
flox	= fl(ocks)ox(en), ciprofloxacin
flu	= flu(id), fluticasone
flur	= flur(ry), sevoflurane
fo	= (ef)fo(rt), tenofovir
fo	= fo(e), fo(am), fosinopril
fol	= fol(licle), fol(ly), propofol
for	= for, for(t), formoterol
fos	= fossil, sofosbuvir
fu	= fu(mes), fu(ry), enfuvirtide (T-20)
fur	= fur(y), nitrofurantoin
ga	= (yo)ga, dabigatran
ga	= ga(s), ga(p), gatifloxacin
gem	= gem, gemfibrozil
gen	= gen(tleman), gentamicin
gens	= (oxy)gens, estrogens
ges	= ges(t), (in)ges(t), levonorgestrel
gi	= (ma)gi(c), (lo)gi(c), selegiline
gine	= (aspara)gine, lamotrigine
glar	= g(old)lar(k), g(ood)lar(d), glargine
gli	= gli(b), gli(nt), glipizide
glip	= (bi)g(s)lip, linagliptin
glu	= glu(ten), glu(e), glucagon
gly	= gly(cemic), polyethylene glycol
gly	= gly(ph), nitroglycerin
gon	= (hexa)gon, gon(e), glucagon
gox	= (oxy)g(en)(b)ox, digoxin
gra	= (ag)gra(vate), raltegravir
gre	= gre(mlin), gre(nade), ticagrelor
grel	= (mon)grel, clopidogrel

How to Pronounce Drug Names

guai	= gua(va)"i", gua(camole)"i", guaifenesin
ha	= ha(t), ha(lf), haloperidol
he	= he(n), he(lp), heparin
hy	= hy(brid), hy(phen), magnesium hydroxide
i	= "i", ibuprofen
i	= i(t), simethicone
ib	= (r)ib, celecoxib
ide	= (sl)ide, magnesium hydroxide
ime	= (ox)ime, cefuroxime
in	= in, (t)in, polymyxin B
ine	= (fam)ine, (eng)ine, levothyroxine
ine	= (sal)ine, desvenlafaxine
ip	= (s)ip, ipratropium
ir	= (f)ir, (s)ir, Irbesartan
ka	= (s)ka(te), Ka(y), amikacin
kast	= (Out)kast, (Di)kast, montelukast
ke	= ke(y), ke(ep), ketoconazole
ki	= ki(d), aliskiren
la	= la(b), la(st), azelastine
la	= (co)la, (Koa)la, desvenlafaxine
la	= la(ser), la(ce), vilazodone
la	= la(wn), la(w), clavulanate
lac	= (li)lac, lac(tose), spironolactone
lam	= lam(b), lam(p), alprazolam
lan	= lan(ce), lan(d), dexlansoprazole
last	= last, roflumilast
late	= (p)late, bismuth subsalicylate
le	= le(an), levothyroxine
le	= le(tter), celecoxib
len	= len(s), len(d), alendronate
lev	= lev(el), (e)lev(ator), levalbuterol
lex	= (f)lex, cephalexin
li	= li(ght), li(e), lidocaine

li	= li(ly), li(nt), palivizumab
lid	= lid, (s)lid, linezolid
lin	= lin(t), amoxicillin
line	= (sa)line, sertraline
line	= line(n), (alka)line [Note: not (fe)line or (sa)line]
lis	= lis(t), lis(ten), lisdexamfetamine
lo	= (Co)lo(rado), (s)lo(w), loperamide
lol	= lol(lipop), lol(lygag), propranolol
lone	= lone, (a)lone, (c)lone, triamcinolone
lor	= lor(d), lor(e), ticagrelor
lox	= lox, meloxicam
lu	= lu(cid), montelukast
ly	= (li)ly, (po)ly(gon), polyethylene glycol
ma	= ma(p), ma(t), temazepam
ma	= ma(ze), may, temazepam
ma	= (paja)ma, somatropin
mab	= m(at)(c)ab, palivizumab
mag	= mag(azine), magnesium hydroxide
man	= man, mannitol
man	= hu(man), memantine
mate	= mate, (room)mate, topiramate
me	= me(n), me(sh), esomeprazole
mel	= mel(on), mel(t), ramelteon
mer	= (ga)mer, mercaptopurine
met	= (co)met, met(al), metformin
meth	= meth(od), simethicone
mi	= (se)mi, (a)mi, amiodarone [Note: rhymes with me]
mi	= mi(le), fidaxomicin
mi	= mi(nt), mi(lk), acetaminophen
mibe	= m(y)(tr)ibe, ezetimibe
mide	= (gu)m(sl)ide, loperamide
mil	= mil(k), verapamil
min	= min(t), (vita)min, metformin
mine	= (hista)mine, brompheniramine
mir	= (ad)mir(al), mir(th), mirtazapine
mir	= mir(ror), detemir
mo	= mo(at), famotidine

mo	= mo(nolith), isosorbide mononitrate
mol	= mol(lusk), (enty)mol(ogy), methocarbamol
mole	= mole, dipyridamole
mon	= mon(key), montelukast
mor	= mor(e), Mor(se)(code), hydromorphone
mox	= mox(ie), amoxicillin
mu	= mu(sic), mu(seum), mupirocin
muth	= (azi)muth, bismuth subsalicylate
my	= my, my(osin), neomycin sulfate
myx	= myx(edema), polymyxin B
na	= na(p), (g)na(t), clonazepam
na	= na(vy), na(me), clonazepam
na	= (bana)na, fexofenadine
nac	= (k)nac(k), diclofenac
nase	= (oxyge)nase, asparaginase
nate	= (in)nate, Nate, calcium carbonate
ne	= (mo)ne(y), k(ne)e, magnesium hydroxide
ne	= ne(t), guaifenesin
ner	= ner(d), etanercept
ni	= (k)ni(t), prednisolone
ni	= ni(ght), brompheniramine
ni	= (mi)ni, aclidinium bromide
nib	= nib(ble), imatinib
nide	= (s)nide, (cya)nide, budesonide
nin	= (mela)nin, oxybutynin
nir	= (souve)nir, cefdinir
no	= no, (k)no(w), phenobarbital
no	= (can)no(n), (ca)no(n), acetaminophen
no	= no(d), no(t), tenofovir
nol	= (etha)nol, (mo)nol(ith), allopurinol
nor	= nor, nor(th), buprenorphine / naloxone
nox	= nox(ious), (equi)nox, diphenoxylate / atropine
ny	= (po)ny, (to)ny, phenytoin
ny	= ny(lon), nystatin
nyl	= (vi)nyl, fentanyl
o	= "o", fexofenadine

ol	= (aw)ol, (alcoh)ol, albuterol
ol	= ol(d), olmesartan
ole	= (m)ole, (h)ole, pramipexole
om	= (m)om, ombitasvir
on	= on, (w)on(ton), ondansetron
one	= (c)one, naloxone
or	= or, (st)or(e), dextromethorphan
ox	= ox, (b)ox, sulfamethoxazole/trimethoprim
pa	= (pa)pa, levodopa
pa	= pa(ir), pa(re), enoxaparin
pa	= pa(th), palivizumab
pam	= pam(phlet), (s)pam, diazepam
pan	= pan, pantoprazole
part	= part, (a)part(ment), insulin aspart
pas	= pas(s), pas(t), ledipasvir
pe	= pe(n), pe(ril), pe(t), loperamide
pen	= pen, pen(ny), gabapentin
per	= per(son), permethrin
pex	= (a)pex, pramipexole
pha	= (al)pha, cephalexin
phan	= (or)phan, dextromethorphan
phe	= phe(nom), phe(notype), phenobarbital
phe	= phe(nomenon), brompheniramine
phen	= (hy)phen, acetaminophen
phine	= (So)phi(a)(du)ne, morphine
pho	= pho(ne), pho(to), amphotericin B
phone	= phone, hydromorphone
phos	= phos(phor), cyclophosphamide
phrine	= phr(ase)(femin)ine, epinephrine
pi	= pi(n), aspirin
pi	= pi(ece), ipratropium
pi	= pi(e), pioglitazone
pime	= pi(ece)(li)me, cefepime
pin	= pin, somatropin
pine	= (shi)p(mar)ine, atropine
pix	= pix(el), pix(ie), Apixaban
pla	= pla(tinum), cisplatin

po	= po(t), po(lygon), polyethylene glycol
po	= po(tato), propofol
pra	= (su)pra, (O)pra(h), dexlansoprazole
pra	= pra(ctice), pra(ttle), alprazolam
pra	= pra(y), pra(ise), alprazolam
pra	= pra(nce), (so)pra(no), propranolol
pram	= pr(e)am(ble), pram, citalopram
pre	= pre(dator), pre(dict), buprenorphine / naloxone
pre	= pre(mie), pre(med), pregabalin
pred	= pred(ator), pred(ict), prednisonolone
pril	= pril(l), benazepril
prim	= prim, prim(p), sulfamethoxazole/trimethoprim
prine	= pr(oton), (mach)ine, cyclobenzaprine
pro	= (a)pro(n), metoprolol
pro	= pro(m), pro(d), lubiprostone
pro	= pro(ton), pro(be), pro(active), ibuprofen
prox	= pr(oton)(b)ox, naproxen
pseu	= pseu(do), pseudoephedrine
pur	= pur(e), pur(ity), allopurinol
py	= py(lon), py(re), pyrazinamide (PZA)
py	= py(xis), dipyridamole
pyl	= (po)p(vin)yl, propylthiouracil
que	= que(ll), que(st), quetiapine
qui	= qui(nce), qui(t), quinapril
quine	= quin(c)e, quin(t)e(t), hydroxychloroquine
ra	= (au)ra, (cob)ra, (Op)ra(h), loperamide
ra	= ra(ttle), loratadine
ra	= ra(y), ra(zor), lorazepam
ral	= ral(ly), Ral(ph), raltegravir
re	= re(d), re(f), varenicline
ren	= (si)ren, ren(t), aliskiren
rene	= (se)rene, triamterene
rex	= (ano)rex(ic), suvorexant
ri	= ri(de), ri(val), rivaroxaban
ri	= ri(nse), (Flo)ri(da), cetirizine
ride	= ride, (p)ride, glyburide

rin	= rin(se), permethrin, [Note: not r-i-n-e]
rine	= (doct)rine, (u)rine, phenylephrine
ro	= ro(w), ro(ad), mupirocin
roc	= roc(k), (c)rock, maraviroc
role	= role, (pa)role, ropinirole
rone	= (d)rone, (p)rone, buspirone
rox	= (xe)rox, p(rox)imal, cefuroxime
ru	= ru(by), ru(in), darunavir
ry	= (o)ry(x), (t)ry(st), erythromycin
sa	= (vi)sa, sa(liva), bismuth subsalicylate
sal	= sal(ad), sal(iva), salmeterol
sar	= sar(dine), (pul)sar, candesartan
sate	= (pul)sate, sate, (compen)sate, docusate sodium
sax	= sax(ophone), sax(ony), saxagliptan
sco	= sco(pe), sco(ne), scopolamine
se	= se(t), se(nt), ondansetron
sel	= sel(fie), sel(l), oseltamivir
ser	= (dre)ser, ser(f), sertraline
si	= (enthu)si(astic), (tran)si(ent), magnesium hydroxide
si	= si(t), simethicone
side	= side, (out)side, etoposide
sil	= sil(ver), sil(l), sildenafil
sim	= sim(ple), sim(mer), simvastatin
sin	= (ba)sin, sin, guaifenesin
so	= so(fa), so, docusate sodium
so	= so(n), prednisolone
sone	= so(fa), (du)ne, fluticasone
sor	= sor(bet), sor(e), isosorbide mononitrate
spa	= spa(re), asparaginase
spe	= spe(ck), spe(nd), risperidone
spi	= spi(ll), spi(n), buspironerone
spo	= spo(rt), cyclosporine
sta	= sta(ff), nystatin
stat	= (photo)stat, stat(im), febuxostat
ste	= ste(ps), ste(ms), testosterone
stine	= (cele)stine, (cy)stine, azelastine
stone	= stone, lubiprostone

su	= su(e), Su(san), sumatriptan
su	= su(n), su(b), insulin
sub	= sub(marine), sub(way), bismuth subsalicylate
suc	= suc(ces), suc(tion), metoprolol succinate
sul	= sul(len), neomycin sulfate
sur	= sur(fer), sur(f), surfactant
ta	= ta(b), ta(x), Oxymetazoline
ta	= (da)ta, (sona)ta, (quo)ta, acetaminophen
ta	= ta(pe), ta(me), mirtazapine
tal	= tal(l), butalbital
tam	= tam(ale), levetiracetam
tan	= (sul)tan, eletriptan
tan	= tan, tan(trum), candesartan
tant	= (sex)tant, surfactant
tar	= tar, tar(get), metoprolol tartrate
tas	= tas(k), potassium
tate	= (es)tate, (mu)tate, benzonatate
tax	= tax, tax(i), metaxalone
te	= te(a), te(e), ramelteon
te	= te(nt), motelukast
tel	= tel(ephone), telmisartan
ter	= ter(se), albuterol
tes	= tes(t), tes(tudo), testosterone
tha	= tha(tch), tha(t), chlorthalidone
thi	= thi(gh), propylthiouracil
thi	= thi(ef), lithium
thro	= thro(w), azithromycin
thy	= thy(roid), thy(mus), levothyroxine
ti	= ti(ara), (zi)ti, levetiracetam
ti	= ti(c), (plas)ti(c), cetirizine
ti	= ti(e), ti(de), tizanidine
tide	= tide, (rip)tide, enfuvirtide (T-20)
tin	= tin, tin(der), nystatin
tine	= (sal)tine, (rou)tine, fluoxetine
to	= (pis)to(n), (a)to(m), ketoconazole
to	= to(e), to(w), pantoprazole
to	= to(t), to(p), testosterone

tol	= tol(erate), (derma)tol(ogy), ethambutol
tone	= tone, (s)tone, nabumetone
tor	= tor(so), tor(t), atorvastatin
tra	= tra(ck), tramadol
tra	= (orches)tra, (spec)tra, (ex)tra, sertraline
tra	= tra(y), bacitracin
tran	= tran(sfer), tran(smit), dabigatran
trate	= (concen)trate, (magis)trate, metoprolol tartrate
trel	= trel(lis), levonorgestrel
trex	= tre(mor), "x", methotrexate
tri	= tri(angle), triamcinolone
tri	= tri(p), tri(ck), clotrimazole
tril	= (nos)tril, tril(l), sacubitril / valsartan
trip	= trip, trip(lets), eletriptan
tro	= (me)tro, (re)tro, dextromethorphan
tron	= (elec)tron, ondansetron
tux	= tux(edo), rituximab
ty	= ty(pical), ty(mpanic), amitriptyline
u	= "u", cefuroxime
um	= (g)um, calcium carbonate
va	= (lar)va, (sali)va, varenicline
vac	= vac(uum), influenza vaccine
val	= val(ley), val(ve), divalproex
van	= van, van(dal), vancomycin
var	= var(sity), var(nish), vardenafil
ve	= ve(rb), (e)ve(r), verapamil
ve	= ve(teran), ve(st), carvedilol
ven	= ven(om), ven(dor), desvenlafaxine
vi	= vi(sion), vi(m), palivizumab
vin	= vin(tage), vincristine
vir	= vir(tue), enfuvirtide
vir	= vir(ulent), oseltamivir [rhymes with veer]
vo	= (di)vo(t), (pi)vo(t), nebivolol
vo	= vo(te), sevoflurane

How to Pronounce Drug Names

vu	= (re)vu(e), clavulanate
war	= war, war(fare), warfarin
y	= (countr)y, oxycodone
y	= (t)y(pical), (h)y(pnotize), hydroxyzine
yl	= (s)yl(lable), polyethylene glycol
za	= (pla)za, influenza vaccine
ze	= (ha)ze(l), clonazepam
ze	= ze(d), ze(phyr), azelastine
zem	= (ec)zem(a), diltiazem [Note: rhymes with them]
zi	= zi(pper), zi(t), azithromycin
zid	= zi(ppe)d, isoniazid(INH)
zide	= z(ipper)(r)ide, hydrochlorothiazide
zil	= (Bra)zil, zil(ch), donepezil
zin	= zin(fandel), canagliflozin
zine	= (maga)zine, cetirizine
zo	= zo(mbie), benzonatate
zo	= (hori)zo(n), trazodone
zo	= zo(ne), (o)zo(ne), (gon)zo, benzocaine
zol	= zo(na)l, zolpidem
zole	= (fe)z(p)ole, sulfamethoxazole/trimethoprim
zone	= zone, (end)zone, pioglitazone
zox	= (qui)z(b)ox, nitazoxanide
zu	= Zu(lu), (shiat)zu, palivizumab

Generic / Brand Index

www.ingramcontent.com/pod-product-compliance
Lightning Source LLC
Chambersburg PA
CBHW032008170526
45157CB00002B/592